T0251420

AIR COOLING TECHNOLOGY
TECHNOLOGY
for ELECTRONIC
EQUIPMENT

About the cover

The inset photograph shows the temperature fields around four one-inch-square simulated electronic components flush mounted  on a conducting board cooled by air flowing from left to right. Isotherms are made visible by using thermochromic liquid crystal thermography. (Photo courtesy of Professor Alfonso Ortego of the Heat Transfer Laboratory at the University of Arizona.)

AIR COOLING TECHNOLOGY for ELECTRONIC EQUIPMENT

Edited by
Sung Jin Kim
Sang Woo Lee

CRC PRESS

Boca Raton London New York Washington, D.C.

Library of Congress Cataloging-in-Publication Data

Air cooling technology for electronic equipment / Sung Jin Kim, Sang Woo Lee, editors
    p. cm.
  Includes bibliographical references and index.
  ISBN 0-8493-9447-3 (alk. paper)
    1. Electronic apparatus and appliances—Cooling. 2. Electronic packaging—Cooling.
  3. Heat—Convection. 4. Air flow I. Kim, Sung Jin. II. Lee, Sang Woo.
TK7870.25.A43 1996
621.381′046—dc20   94-45903
                                                               CIP

Visit the CRC Press Web site at www.crcpress.com

FOREWORD

Cooling technology has been a vital prerequisite for the rapid, if not explosive, growth of the electronic equipment industry. This has been especially true during the last 15 years with the advent of integrated circuit chips and their applications in computers and related electronic products. Today, cooling technology is well recognized as a gating factor for the development of future electronic equipment. There are many reasons why cooling technology always will be pivotal in the design of electronic equipment, but the following two are most fundamental:

- Thermodynamically, all electronic devices are undergoing an irreversible process with the net result being the generation of heat which must be removed in order to maintain continuous operation.

- Thermophysically, the reliability and performance of all electronic devices are temperature dependent with various degrees of sensitivity, but generally the lower the temperature, the better.

With these in mind, the search for new and improved cooling technologies has been unabated for the last three decades as evidenced by both industrial development efforts and university research activities. IBM has been involved actively with sponsored electronic cooling research at many universities across the country since 1967 with equal emphasis on both air cooling and liquid cooling. Almost without exception, all research findings sponsored by IBM have been published, contributing substantially to the advancement of cooling technology for the industry.

Recent developments since 1992 have resulted in adapting CMOS technology for pervasive applications in almost all commercial electronics, including computers. This development, coupled with the increasingly competitive nature of the industry has made air cooling much more enticing. Therefore, a book on "Air Cooling Technology for Electronic Equipment" is both natural and logical, and I endorse this timely idea wholeheartedly. Dr. Sung Kim and Dr. Sang Lee, co-editors of this book, are my colleagues at IBM and all chapter authors are my professional friends of many years. I have every reason to believe that their joint effort will be a successful one.

Richard C. Chu
IBM Fellow

PREFACE

Improvements in microelectronics and large scale integration technologies result in power densities as high as 10 W/cm^2 for semiconductor devices. This continuous increase in power densities has placed an increased emphasis on the thermal control of electronic equipment. Convection air cooling is still the most common method for removing heat from the heat generating elements. This is because air is readily available and the air moving devices are relatively inexpensive. Even though there are a lot of research activities on "exotic" and "expensive" cooling technologies, more and more emphasis is placed on extending the limits of air cooling capabilities because of the competitive nature of the computer industry.

The following questions arise with regard to the thermal design of electronic systems. What is the optimal spacing between the printed circuit boards? What is a good estimate of the heat transfer coefficient and the associated pressure drop for forced convection over package arrays? How are heat transfer and fluid flow characteristics in the entrance region different from those in the fully developed region? What is the effect of substrate conduction on convection cooling? How can heat transfer be enhanced to push the limit of air cooling? What is the eventual upper limit of air cooling? These questions are frequently asked by thermal engineers and packaging practitioners.

This book addresses all these questions in detail. Six chapters are designated to answer the above-mentioned questions systematically. They are based on research projects funded primarily by IBM. In addition, we have tried to include the most recent information on air cooling technologies performed outside of IBM. Hence, this book will serve as a handy technical source of information for thermal and packaging engineers who would like to get the most out of air cooling. The book in its present state reflects the latest development in cooling techniques and thermal design guides with air as a cooling medium.

It goes without saying that this book would not be possible without the dedication of the chapter authors. We would like to extend our thanks to Bob Stern and Norm Stanton of CRC Press for their enthusiastic support. Our greatest thanks go to our wives, Yeon S. Kim and Ann H. Lee, for their continuous prayers and full support during the last two years. In addition, we would particularly like to thank Mohinder Grewal and Ron Russell of IBM Corporation for providing a working environment that makes a fruitful result such as this possible. Special thanks to Mr. Richard Chu, IBM Fellow, for the insightful foreword.

Sung J. Kim
Sang W. Lee

CONTRIBUTORS

Yutaka Asako
Professor of Mechanical
  Engineering
Tokyo Metropolitan University
Tokyo, Japan

Kaveh Azar
Member of Technical Staff
AT&T Bell Laboratories
North Andover, MA

Adrian Bejan
J.A. Jones Professor
  of Mechanical Engineering
Duke University
Durham, NC

Mohammad Faghri
Professor of Mechanical
  Engineering
University of Rhode Island
Kingston, RI

Suresh Garimella
Cray Research Associate Professor
  of Mechanical Engineering
University of
  Wisconsin–Milwaukee
Milwaukee, WI

Sung Jin Kim
Advisory Engineer
  Thermal Engineering Center
  IBM Storage Systems Division
Tucson, AZ

Sang Woo Lee
Adjunct Professor
  of Mechanical Engineering
Duke University
  Durham, NC

Majid Molki
Professor of Mechanical
  Engineering
Esfahan University of Technology
Esfahan, Iran

Alfonso Ortega
Associate Professor
Department of Aerospace
  & Mechanical Engineering
University of Arizona
Tucson, AZ

Richard Wirtz
Professor of Mechanical Engineering
University of Nevada
Reno, NV

TABLE OF CONTENTS

Chapter 1

GEOMETRIC OPTIMIZATION OF COOLING TECHNIQUES

Adrian Bejan

CONTENTS

0-8493-9447-3/96/$0.00+$.50
© 1996 by CRC Press, Inc.

1

INTRODUCTION

In the design of packages of electronic components there are strong incentives to mount as much circuitry as possible in a given space. This can be achieved by judiciously selecting the *geometry* of the package, i.e., the way in which the components are arranged relative to the coolant and to each other in the fixed space. An important constraint is that the highest temperature (the "hot spot") that is registered at a certain point in the package must not exceed a specified ceiling value. If the temperature rises above the allowable limit, the error-free operation of the electronic circuit is threatened. Since each component in the package generates heat, this design objective translates into maximizing the total rate of heat transfer from the finite space occupied by the package to the coolant that flows through the package.

In the electronics industry there is a great diversity of components, packages, and cooling techniques [Moffat and Ortega, 1988; Peterson and Ortega, 1990; Ishizuka, 1992, 1993]. Because of this diversity each optimal cooling arrangement that emerges out of the design process tends to be specific to a single application and lacks general applicability. The challenge is to identify in this great diversity of configurations those design optimization rules that can be applied to one or more classes of package configurations. In earlier reviews [Bejan and Lee, 1994; Bejan, 1995a] we identified several of the fundamental ways in which the geometric features of entire classes of electronic packages can be optimized. In this chapter, I systematically review the progress made in the newly emerging field of *Geometric Optimization of Cooling Techniques*.

STACK OF VERTICAL PLATES COOLED
BY NATURAL CONVECTION

A sufficiently large number of parallel electronic circuit boards must be cooled by natural convection in the space of height H, thickness L, and width W shown in Figure 1. The width is perpendicular to the plane of the figure, and the board thickness is negligible. Cold fluid (e.g., air) of temperature T_∞ enters through the bottom of the package, rises through the board-to-board channels, and exits through the upper opening. The lateral walls of area H × W that confine the package are insulated. The ceiling value of the board temperature, T_{max} is set by electronic operational constraints.

The objective of the design is to maximize the total heat transfer rate removed by the coolant from the H × L × W space. The boards are assumed equidistant. The only variable is the number of boards

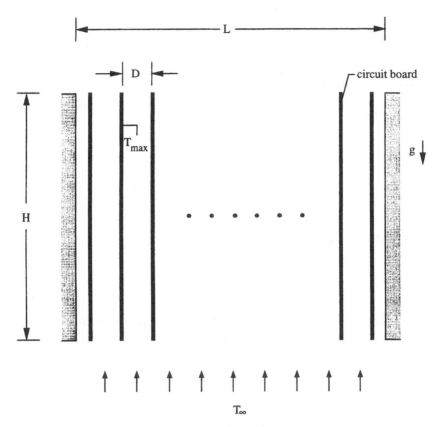

FIGURE 1. Vertical boards cooled by natural convection (Bejan, 1984, 1995a).

$$n \cong \frac{L}{D} \tag{1}$$

or the board-to-board spacing D. To determine the optimal D that maximizes $\dot{Q}$ can be a laborious task, however, we will show that it is possible to solve the problem almost on the back of an envelope. *The intersection of asymptotes method* was outlined first by Bejan [1984], and begins with the assumptions that:

1. The flow is laminar.
2. The board surfaces are sufficiently smooth to justify the use of heat transfer results for natural convection over vertical smooth walls.
3. The maximum temperature T_{max} is, in an order of magnitude sense, representative of the temperature at every point on the board surface.

The method consists of two steps. In the first step, we identify the two extremes in which the cooling process may function, the small-D limit and the large-D limit. In the second step, the two extreme regimes are intersected for the purpose of locating the D value that maximizes $\dot{Q}$.

THE SMALL-D ASYMPTOTE

Consider first the limit of vanishingly small board-to-board spacing, $D \rightarrow 0$. In this limit we can use with confidence $\dot{Q}_1 = \dot{m}_1 c_p (T_{max} - T_\infty)$ for the heat transfer rate extracted by the coolant from one of the channels of spacing D. Note that in each channel T_∞ is the inlet temperature, T_{max} is the outlet temperature, and $\dot{m}_1$ is the mass flow rate through a single channel [Bejan, 1984]

$$U = \frac{g\beta(T_{max} - T_\infty)D^2}{12v} \qquad (2)$$

$$\dot{m}_1 = \rho \, DWU = \frac{\rho W \, g\beta(T_{max} - T_\infty)D^3}{12v} \qquad (3)$$

The total rate of heat transfer removed from the package is $\dot{Q} = n \, \dot{Q}_1$, or

$$\dot{Q} = \rho c_p \, WL \frac{g\beta(T_{max} - T_\infty)^2 D^2}{12v} \qquad (4)$$

In conclusion, in the $D \rightarrow 0$ limit the total heat transfer rate decreases as D^2. This trend is indicated by the small-D asymptote plotted in Figure 2.

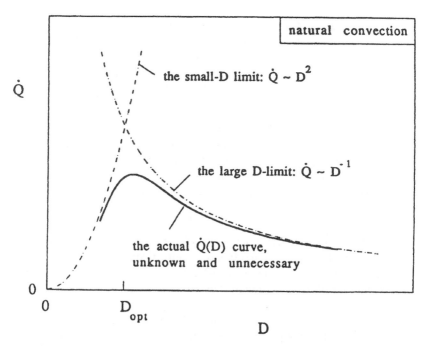

FIGURE 2.   The maximization of the total heat transfer rate removed by natural convection from the vertical stack of Figure 1.

THE LARGE-D ASYMPTOTE

Consider next the opposite limit in which D is large enough so that it exceeds the thickness of the thermal boundary layer that forms on each vertical surface, namely [Bejan, 1995a]

$$D > H \left[\frac{g\beta \, H^3 (T_{max} - T_\infty)}{\alpha v} \right]^{-1/4} \tag{5}$$

if the Prandtl number is of order 1 or greater. In this limit the boundary layers are distinct (i.e., thin compared with D), and the center region of the board-to-board spacing is occupied by fluid of temperature T_∞. The number of distinct boundary layers is $2n = 2L/D$, because there are two boundary layers for each D spacing. The heat transfer rate through one boundary layer is $\overline{h}$ HW $(T_{max} - T_\infty)$ for which $\overline{h}$ is furnished by the correlation for laminar flow only [Bejan, 1993]

$$\frac{\overline{h} \, H}{k} = 0.517 \, Ra_H^{1/4} \tag{6}$$

where $Ra_H = g\beta \, (T_{max} - T_\infty) \, H^3/(\alpha v)$. The total rate of heat transfer extracted from the entire package is 2n times larger than $\overline{h}$ HW $(T_{max} - T_\infty)$

$$\dot{Q} = 2 \frac{L}{D} HW (T_{max} - T_\infty) \frac{k}{H} \, 0.517 \, Ra_H^{1/4} \tag{7}$$

Equation 7 shows that in the large-D limit the total heat transfer rate decreases as D^{-1} as the board-to-board spacing increases. This second asymptote has also been plotted in Figure 2.

THE OPTIMAL PLATE-TO-PLATE SPACING

What we have determined so far are the two asymptotes of the actual (unknown) curve of $\dot{Q}$ versus D. Figure 2 shows that the asymptotes intersect above what would be the peak of the actual $\dot{Q}$ (D) curve. It is not necessary to know the actual $\dot{Q}$ (D) relation. The optimal spacing D_{opt} for maximum $\dot{Q}$ can be estimated (approximately) as the D value where Equations 4 and 7 intersect [Bejan, 1984]:

$$\frac{D_{opt}}{H} \cong 2.3 \left[\frac{g\beta (T_{max} - T_\infty) H^3}{\alpha v} \right]^{-1/4} \tag{8}$$

This estimate reproduces within 20% the optimal spacing deduced based on more exact and lengthier methods, such as the maximization of the $\dot{Q}$(D) relation [Bar-Cohen and Rohsenow, 1984] and the finite-difference simulations of the complete flow and temperature fields in the package [Anand et al., 1992].

An order of magnitude estimate for the maximum heat transfer rate can be obtained by substituting D_{opt} in Equations 7 or 4

$$\dot{Q}_{max} \lesssim 0.45\, k\,(T_{max} - T_\infty)\frac{LW}{H}\, Ra_H^{1/2} \tag{9}$$

The unequal sign is a reminder that the peak of the actual $\dot{Q}(D)$ curve falls under the intersection of the two asymptotes (Figure 2); however, the right-hand side of Equation 9 represents the appropriate scale of the maximum heat transfer rate. This result can be expressed also as the maximum volumetric rate of heat generation in the $H \times L \times W$ space

$$\frac{\dot{Q}_{max}}{HLW} \lesssim 0.45\, \frac{k}{H^2}\,(T_{max} - T_\infty)\, Ra_H^{1/2} \tag{10}$$

In conclusion, if the heat transfer mechanism is natural convection, the maximum density of heat generating electronics (or $\dot{Q}/HLW$) is proportional to $(T_{max} - T_\infty)^{3/2}$ and $H^{-1/2}$. The same figure of merit is proportional to the property group $k(g\beta/\alpha v)^{1/2}$.

BUNDLE OF HORIZONTAL CYLINDERS (PIN FINS) COOLED BY NATURAL CONVECTION

An optimal spacing for natural convection cooling exists regardless of the shape of the heating elements that occupy the $H \times L$ cross-section of the fixed volume of Figure 1. In the preceding section we illustrated the method of intersecting the asymptotes by using the vertical board as the elementary geometric feature. Now we focus on the case where the volume is occupied by an array of horizontal cylinders of diameter D and length W, for example, by pin fins on a vertical base surface $H \times L$, Figure 3 [Bejan et al., 1995a].

When the cylinders are arranged in an equilateral triangular array, the intersection of the large-S and small-S asymptotes yields an implicit relation between the optimal spacing S_{opt} and the Rayleigh number $Ra_D = g\beta D^3 (T_W - T_\infty)/(\alpha v)$

$$\frac{S_{opt}}{D} \cdot \frac{2 + S_{opt}/D}{(1 + S_{opt}/D)^{2/3}} \cong 2.75 \left(\frac{H}{D}\right)^{1/3} Ra_D^{-1/4} \tag{11}$$

This relation is plotted in Figure 4, which shows that S_{opt}/D is almost proportional to the group $(H/D)^{1/3}\, Ra_D^{-1/4}$. In other words, a simpler alternative to the order of magnitude estimate obtained in Equation 11 is

$$\frac{S_{opt}}{D} \sim \left(\frac{H}{D}\right)^{1/3} Ra_D^{-1/4} \tag{12}$$

or

$$\frac{S_{opt}}{H} \sim \left(\frac{H}{D}\right)^{1/12} Ra_H^{-1/4} \tag{13}$$

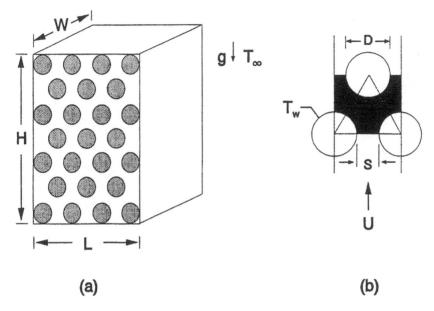

FIGURE 3. Array of horizontal cylinders cooled by natural convection and detail of one of the channels traveled by the coolant (Bejan et al., 1995a).

where $Ra_H = g\beta H^3(T_w - T_\infty)/(\alpha v)$. Equation 13 shows that the optimal spacing is approximately proportional to $H^{1/3}D^{-1/12}$, which means that it is almost insensitive to changes in the cylinder diameter.

The optimal spacing was verified and calculated more accurately based on full numerical simulations and experimental measurements (Bejan et al., 1995a). The numerical results are shown in Figure 4. They are correlated very nicely in the manner anticipated in Equation 11, namely by using the group $(H/D)^{1/3}$ $Ra_D^{-1/4}$ on the abscissa. The optimal spacings, however, are consistently 2.5 times larger than the values calculated based on Equation 11. The function of type 12 that fits the numerical data the best (within 1.7% mean error) is

$$\frac{S_{opt}}{D} = 2.72 \left(\frac{H}{D}\right)^{1/3} Ra_D^{-1/4} + 0.263 \qquad (14)$$

where 0.263 is a small correction term. The numerical results for the maximum heat transfer rate that corresponds to the optimal spacings of Figure 4 are correlated within 1.7% by the expression

$$\frac{\dot{Q}_{max} D^2}{HLWk(T_w - T_\infty)} = 0.448 \left[\left(\frac{H}{D}\right)^{1/3} Ra_D^{-1/4}\right]^{-1.6} \qquad (15)$$

In this expression $\dot{Q}_{max}$ is the total heat transfer removed by natural convection from the fixed volume $H \times L \times W$. Experiments conducted in the range $300 < Ra_D < 400$, $H/D = 6.2$ and $Pr = 0.72$, revealed optimal spacings

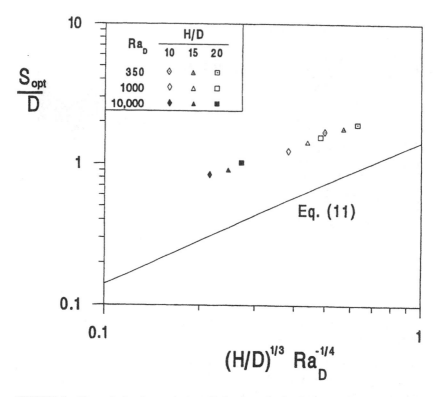

FIGURE 4. Theoretical and numerical results for the optimal cylinder-to-cylinder spacing in natural convection (Bejan et al., 1995a).

that agree within 17% with the values determined numerically. It was also shown that the optimal spacing is relatively insensitive to whether the cylinders are isothermal or with uniform heat flux. This conclusion widens the applicability of Equation 14 and agrees with a similar conclusion reached in Bejan and Sciubba [1992] for stacks of parallel boards cooled by forced convection.

STACK OF PARALLEL PLATES COOLED BY LAMINAR FORCED CONVECTION

Consider now the problem of installing the optimal number of heat generating boards in a space cooled by forced convection [Bejan and Sciubba, 1992]. As shown in Figure 5, the swept length of each board is L, while the transversal dimension of the entire package is H. The width of the stack, W, is perpendicular to the plane of the figure. We continue to rely on the assumptions (a) to (c) listed under Equation 1. The thickness of the individual board is again negligible relative to the board-to-board spacing D, so that the number of boards is approximately

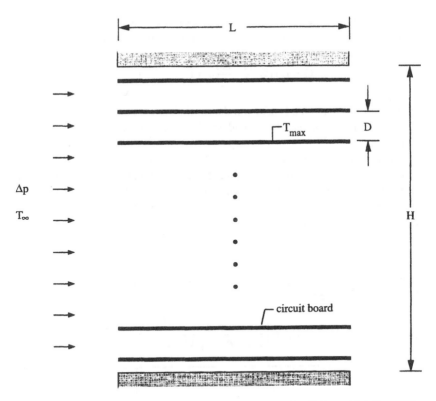

FIGURE 5. Package of parallel boards cooled by forced convection (Bejan and Sciubba, 1992).

$$n \cong \frac{H}{D} \tag{16}$$

It is important that the pressure difference across the package, Δp, is constant and known. This is a good model for electronic systems in which several packages and other features (e.g., channels) receive their coolant in parallel from the same plenum. The plenum pressure is maintained by a fan, which may be located upstream or downstream of the package.

We begin with the class of applications where the flow is laminar. The extension to turbulent flow is discussed in a later section. When D becomes sufficiently small, the channel formed between two boards becomes slender enough for the flow and heat transfer to be in the fully developed regime. The average longitudinal velocity through the channel is [e.g., Bejan, 1993]

$$U = \frac{D^2}{12\mu} \frac{\Delta p}{L} \tag{17}$$

The total mass flow rate through the package of frontal area $H \times W$ is $\dot{m} = \rho H W U$. In the same limit, the mean outlet temperature of the fluid approaches the board temperature T_{max}. The total rate of heat transfer removal from the $H \times L \times W$ space is $\dot{Q} = \dot{m} c_p (T_{max} - T_\infty)$ or, in view of Equation 17

$$\dot{Q} = \rho\, HW \frac{D^2}{12\mu} \frac{\Delta p}{L} c_p (T_{max} - T_\infty) \qquad (18)$$

In this way we reach the conclusion that in the $D \to 0$ limit the total heat transfer rate decreases as D^2. This trend is illustrated by the small-D asymptote in Figure 6.

Distinct thermal boundary layers will cover the board surfaces when the spacing D becomes sufficiently large. Since the overall pressure drop Δp is fixed, the immediate question is what free stream velocity U_∞ sweeps these boundary layers? We answer this question by noting the force balance on the control volume $H \times L \times W$

$$\Delta p \cdot HW = \bar{\tau} \cdot 2n\, LW \qquad (19)$$

in which $\bar{\tau}$ is the L-averaged wall shear stress provided by the Blasius solution [Bejan, 1993]. The result for U_∞ is

$$U_\infty = \left(\frac{\Delta p\, H}{1.328\, n\, L^{1/2} \rho\, v^{1/2}} \right)^{2/3} \qquad (20)$$

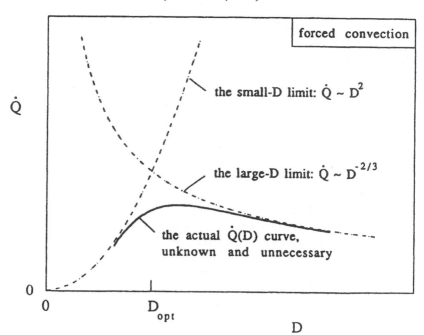

FIGURE 6.   The maximization of the total heat transfer rate removed by forced convection from the package of Figure 5 (Bejan and Sciubba, 1992).

The heat transfer rate through a single board surface is $\dot{Q}_1 = \bar{h}\,LW\,(T_{max} - T_\infty)$, for which the L-averaged heat transfer coefficient $\bar{h}$ is provided by the Pohlhausen solution for $Pr \gtrsim 0.5$ [Bejan, 1993]. The total heat transfer rate removed from the entire package is $\dot{Q} = 2n\,\dot{Q}_1$ or, after using the U_∞ expression listed above

$$\dot{Q} = 1.208\,k\,HW\,(T_{max} - T_\infty)\left(\frac{Pr\,L\,\Delta p}{\rho\,v^2 D^2}\right)^{1/3} \tag{21}$$

In conclusion, in the large-D limit the total heat transfer rate decreases as $D^{-2/3}$ while the board-to-board spacing increases. This second trend is also visible in Figure 6.

The intersection of the two $\dot{Q}(D)$ asymptotes, Equations 18 and 21, yields an approximate estimate for the optimal board-to-board spacing for maximum heat transfer rate

$$\frac{D_{opt}}{L} \cong 2.7\left(\frac{\Delta p \cdot L^2}{\mu\alpha}\right)^{-1/4} \tag{22}$$

This optimal spacing increases as $L^{1/2}$, and decreases as $\Delta p^{-1/4}$ if the applied pressure difference increases. The D_{opt} estimate produced by Equation 22 agrees within 20% with the more exact value obtained by locating the maximum of the actual $\dot{Q}(D)$ curve [Bejan and Sciubba, 1992] and is adequate when the board surface is modeled as isothermal. For surfaces modeled as uniform flux, the 2.7 coefficient is replaced by 3.2 on the right side of Equation 22.

It is instructive to compare Equation 22 with Equation 8 and to notice the symmetry between the nondimensional optimal spacings for forced convection cooling and natural convection cooling. Note further that the role in the natural convection formula is played by Ra_H, in forced convection is played by the new dimensionless pressure drop number [Bejan, 1993]:

$$\Pi_L = \frac{\Delta p \cdot L^2}{\mu\alpha} \tag{23}$$

This dimensionless group was also identified in problems of external forced convection [Bhattacharjee and Grosshandler, 1988], entropy generation minimization [Paoletti et al., 1989], and electronic cooling [Knight et al., 1991; Bejan and Sciubba, 1992] as summarized in Petrescu [1994].

The manner in which the design parameters influence the maximum rate of heat removal from the package is indicated by

$$\dot{Q}_{max} \lesssim 0.6\,k\,(T_{max} - T_\infty)\frac{HW}{L}\,\Pi_L^{1/2} \tag{24}$$

which is obtained by setting $D = D_{opt}$ in Equation 18 or 21. Once again, the unequal sign is a reminder that the actual $\dot{Q}_{max}$ may be smaller by a few tens of percentage points, because the peak of the $\dot{Q}(D)$ curve is situated under the point where the two asymptotes cross in Figure 6. The maximum volumetric rate of heat generation in the $H \times L \times W$ space is

$$\frac{\dot{Q}_{max}}{HLW} \lesssim 0.6 \frac{k}{L^2}(T_{max} - T_\infty)\Pi_L^{1/2} \tag{25}$$

In a subsequent study [Mereu et al., 1993] it was shown that the optimal board-to-board spacing (Equation 22) also holds when the board thickness (t) is not negligible in comparison with the spacing (D). This conclusion was based on the method of the intersection of the asymptotes, as well as on full numerical simulations of the flow field upstream, inside, and downstream of the stack. The maximum overall thermal conductance between the coolant (T_∞ at inlet) and the entire stack (hot-spot temperature T_{max}, total heat generation rate $\dot{Q}_{max}$) is correlated within 5% by the formula

$$\frac{\dot{Q}_{max}/[W(T_{max} - T_\infty)]}{Hc_p(\rho\Delta p/Pr)^{1/2}} = \frac{0.18}{1 + 0.36(t/L)\Pi_L^{1/4}} \tag{26}$$

This correlation was tested numerically in the Re range 500–1000, where Re is based on the board length L and the approach velocity of the free stream. It is relatively insensitive to whether the surfaces are isothermal or with uniform flux. Again, the optimization of the D spacing was based on the assumption that the overall pressure drop across the stack is fixed. Corresponding optimal D and maximum conductance results have also been developed for stacks subjected to fixed total mass flow rate and fixed pumping power [Mereu et al., 1993].

STACK OF PARALLEL PLATES WITH FLUSH-MOUNTED AND PROTRUDING HEAT SOURCES

The optimization of the spacing between boards with discrete heat sources was performed numerically [Morega and Bejan, 1994a]. This was an extensive, systematic study where both flush-mounted and protruding sources were considered. A sample of the optimization procedure is shown in Figure 7, which corresponds to a stack where five flush-mounted heat sources are energized on only one side of each board. The other side of the board is assumed adiabatic. In addition to the spacing D and swept length L, the geometry of each channel is complicated by the source length L_s, unheated length L_d, and total heated length ℓ. The heat generation rate is the same in every heat source. The hot-spot temperature (T_{max}) occurs at the trailing edge of the last heat source. In dimensionless form, the maximum overall thermal conductance of the stack (B_{max}) is the group on the left side of Equation 26,

$$B_{max} = \frac{\dot{Q}_{max}/[W(T_{max} - T_\infty)]}{Hc_p(\rho\Delta p/Pr)^{1/2}} \tag{27}$$

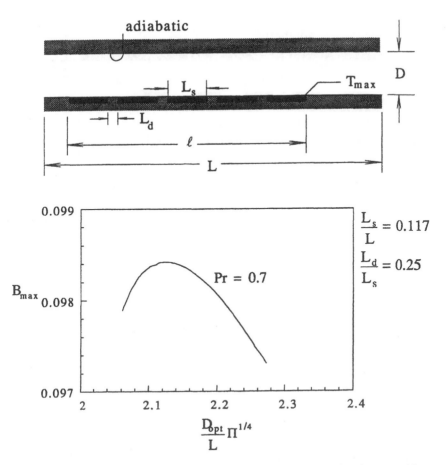

FIGURE 7. The maximization of the overall thermal conductance when heat is generated by five equidistant flush-mounted heat sources (Morega and Bejan, 1994).

Another example of the optimization procedure is presented in Figure 8. When the heat sources are raised, the geometric description is further complicated by the source height D_s. The abscissas of Figures 7 and 8 show that the scaling law (Equation 22) can be used to predict adequately the optimal spacing. Many geometries of these types (Figure 7 and 8) were optimized, and the conclusion was that although Equation 22 works, a more accurate correlation is the one based on the actual heated length (ℓ) instead of L:

$$\frac{D_{opt}}{\ell} \cong 2.7 \, \Pi_\ell^{-1/4} \tag{28}$$

The effect of fluid type (Pr) is already built into Equation 28 by using the pressure drop number defined in Equation 23. The overall thermal conductance of

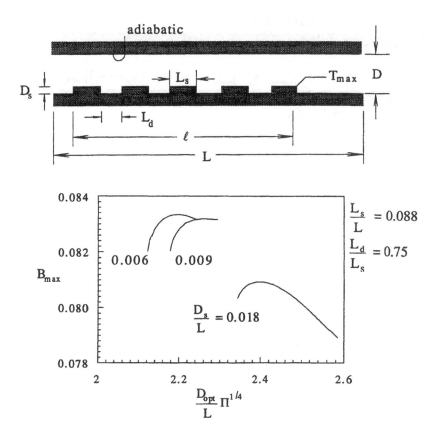

FIGURE 8.   The effect of varying the height of the protruding heat sources in a stack of parallel boards cooled by forced convection (Morega and Bejan, 1994a).

the stack is correlated adequately by $B_{max} \sim 0.08$, which in physical terms means that

$$\frac{\dot{Q}_{max} / W}{T_{max} - T_\infty} \sim 0.08\, k \frac{H}{L} \Pi_L^{1/2} \tag{29}$$

Although the various changes in geometry and thermal boundary conditions have some effect, the actual D_{opt} and $\dot{Q}_{max}/(T_{max} - T_\infty)$ values are relatively insensitive to these details and, for engineering purposes, are described quite adequately by Equations 28 and 29. Furthermore, the peak of the thermal conductance curve is sufficiently flat so that it is not absolutely necessary to know the precise value of D_{opt}. The numerical results also showed that if discrete heat sources are mounted on both sides of each board, the D_{opt} value is roughly 30% greater than the estimate based on Equation 28 and that the maximum conductance is about 3 times greater than the value furnished by Equation

29. Boards with protruding heat sources and laminar forced convection were analyzed numerically in Kim and Anand [1994a] and numerically and experimentally in Farhanieh et al. [1993].

STACK OF PARALLEL PLATES COOLED BY TURBULENT FORCED CONVECTION

The optimal spacing of boards with turbulent flow has been determined based on the method of intersecting the asymptotes [Bejan and Morega, 1994]

$$\frac{D_{opt}/L}{(1+t/D_{opt})^{1/2}} = (f\,C_f)^{1/2}\,Pr^{-2/3} \qquad (Pr \lesssim 0.5) \tag{30}$$

In this expression t, f, and C_f are the board thickness, the friction factor for fully developed flow through a parallel-plate channel, and the average skin friction coefficient for a plane surface, smooth or rough. The corresponding maximum value of the overall thermal conductance is

$$\left[\frac{q'L}{kH(T_{max}-T_\infty)}\right]_{max} \leq \left(\frac{C_f}{f}\right)^{1/4} Pr^{1/6}\left(1+\frac{t}{D_{opt}}\right)^{-3/4}\left(\frac{\Delta p \cdot L^2}{\mu\alpha}\right)^{1/2} \tag{31}$$

$(Pr \gtrsim 0.5)$

where $q' = \dot{Q}/W$. The inequality sign is a reminder that if q' is plotted on the ordinate and D on the abscissa, the peak of the actual q' vs. D curve is located under the intersection of the asymptotes (Figure 6). The right side of Equation 31 represents the correct order of magnitude of the maximum overall thermal conductance and can be expected to anticipate within 30% the exact value.

Beyond this point we must make certain assumptions regarding the values of the friction factor and skin-friction coefficient. If all the board surfaces are smooth, we can use the standard correlations [e.g., Bejan, 1993]

$$f = 0.046\,Re_{D_h}^{-1/5} \qquad (10^4 < Re_{D_h} < 10^6) \tag{32}$$

$$\frac{1}{2}C_f = 0.037\,Re_L^{-1/5} \qquad (10^6 < Re_L < 10^8) \tag{33}$$

where $D_h = 2D$, $Re_{D_h} = D_h U/\nu$ and $Re_L = U_\infty L/\nu$. These allowed us to relate U and U_∞ to Δp as shown in Bejan and Morega [1994]:

$$U = 5.98\,D^{2/3}\nu^{-1/9}\left(\frac{\Delta p}{\rho L}\right)^{5/9} \tag{34}$$

$$U_\infty = 4.25 \, L^{-4/9} v^{-1/9} \left[\frac{\Delta p \, (D + t)}{\rho} \right]^{5/9} \qquad (35)$$

When combined, Equations 32–35 express f and C_f as functions of the imposed pressure drop, i.e., functions that can be substituted on the right side of Equation 30. The final expression for the optimal spacing is

$$\frac{D_{opt}/L}{(1 + t/D_{opt})^{4/11}} = 0.071 \, Pr^{-5/11} \left(\frac{\Delta p \cdot L^2}{\mu \alpha} \right)^{-1/11} \qquad (36)$$

The geometric meaning of this conclusion becomes clearer if we estimate the expected order of magnitude of the right side of Equation 36. First, note that the Re_{DH} range listed in Equation 32 can be rewritten in terms of Δp by using Equation 34 and the assumptions that $(1 + t/D_{opt})^{4/11} \cong 1$ and $Pr = 0.72$ (air)

$$0.09 > \left(\frac{\Delta p \cdot L^2}{\mu \alpha} \right)^{-1/11} > 0.032 \qquad (37)$$

Similarly, the Re_L range specified in Equation 33 can be rewritten using Equation 35

$$0.087 > \left(\frac{\Delta p \cdot L^2}{\mu \alpha} \right)^{-1/11} > 0.038 \qquad (38)$$

Equations 37 and 38 show that the specified Re_{DH} and Re_L ranges correspond to the same range of the pressure drop number $\Delta p \cdot L^2/(\mu \alpha)$. Taken together, Equations 36–38 show that the slenderness ratio of each board-to-board channel (D_{opt}/L) takes values between approximately 0.003 and 0.007 and is relatively insensitive to changes in the applied pressure difference.

When the surfaces are smooth, cf. Equations 32 and 33, the maximum overall conductance expression (Equation 31) becomes

$$\left[\frac{q'L}{kH \, (T_{max} - T_\infty)} \right]_{max} \leq 0.57 \, Pr^{4/99} \left(1 + \frac{t}{D_{opt}} \right)^{-67/99} \left(\frac{\Delta p \cdot L^2}{\mu \alpha} \right)^{47/99} \qquad (39)$$

where $q' = \dot{Q}/W$. In the case of a fluid with Prandtl number of order 1, Equation 39 is almost the same as the more general Equation 31 with the constant 0.57 in place of $(C_f/f)^{1/4}$. In conclusion, the maximum overall conductance increases almost as $\Delta p^{1/2}$.

Figure 9 shows the optimal spacing calculated by using Equation 36 for turbulent flow. The corresponding D_{opt}/L result for laminar flow has been plotted to the left while using the coefficient 3.2 (uniform flux) instead of 2.7 (uniform temperature) on the right side of Equation 22. The figure shows that when the flow is turbulent D_{opt}/L depends not only on the pressure drop number $\Delta p \cdot L^2/(\mu \alpha)$ but also on Pr and t/L. The optimal spacing in turbulent flow increases as Pr and t/L increase and is quite sensitive to such changes.

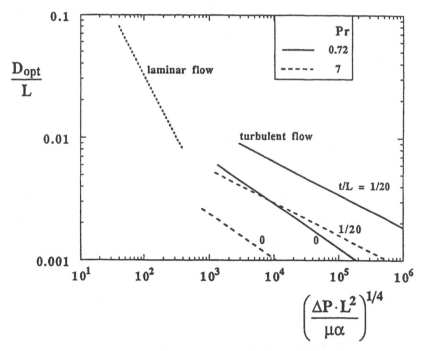

FIGURE 9. The optimal board-to-board spacing for laminar and turbulent forced convection as a function of the pressure drop number, the Prandtl number, and the board slenderness ratio t/L (Bejan and Morega, 1994).

The corresponding maximum overall thermal conductance for turbulent flow (Equation 39) has been plotted in Figure 10 next to the results for laminar flow. The interesting conclusion made possible by this figure is that the turbulent portion of each curve is, in an order of magnitude sense, an extension of the laminar portion. This feature is unlike Figure 9 where there is a sharp change in the behavior of D_{opt}/L as the flow regime changes.

For boards with rough surfaces, the optimal spacing can be calculated with Equations 30 and 31 and appropriate correlations for f and C_f. As a start, it can be noted that when the roughness features are large (e.g., in the fully rough turbulent regime in a duct [Bejan 1993]) the friction factor and the skin friction coefficient are relatively constant, i.e., insensitive to the Reynolds number. Boards with protruding heat sources cooled by turbulent forced convection were studied numerically by Kim and Anand [1994b] and experimentally in Li et al. [1993].

STACK OF PARALLEL PLATES IMMERSED IN A FREE STREAM

Another way of using the information of Figures 9 and 10 is to consider a cooling arrangement in which specified is not Δp but the coolant velocity well upstream of the stack, U_0. In such an arrangement, the scale of Δp across each channel has the following order of magnitude [Morega and Bejan, 1994a]

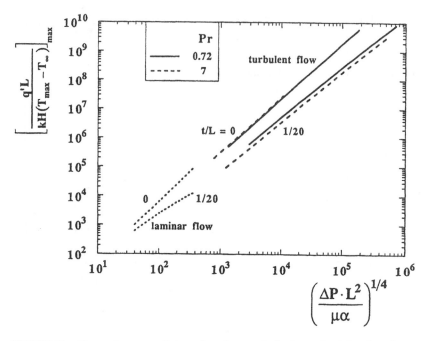

FIGURE 10. The maximum overall thermal conductance for laminar and turbulent forced convection as a function of the pressure drop number, the Prandtl number, and the board slenderness ratio t/L (Bejan and Morega, 1994).

$$\Delta p \cong \frac{1}{2} \rho U_0^2 \tag{40}$$

This scale can be substituted in Equations 36 and 22 to obtain [Bejan and Morega, 1994]

$$\frac{D_{opt}/L}{(1 + t/D_{opt})^{4/11}} \cong 0.076 \, Pr^{-6/11} \, Re_L^{-2/11} \quad \text{(turbulent)} \tag{41}$$

$$\frac{D_{opt}}{L} \cong 3.8 \, Pr^{-1/4} \, Re_L^{-1/2} \quad \text{(laminar)} \tag{42}$$

The Reynolds number Re_L is based on the specified upstream velocity and the flow length of the stack

$$Re_L = \frac{U_0 L}{\nu} \tag{43}$$

The optimal spacings recommended by Equations 41 and 42 are displayed in Figure 11, which shows that in turbulent flow the channel spacing is influenced not only by Re_L but also by Pr and t/L.

The maximum overall thermal conductance (Equation 39) can also be expressed in terms of Re_L by using Equation 40

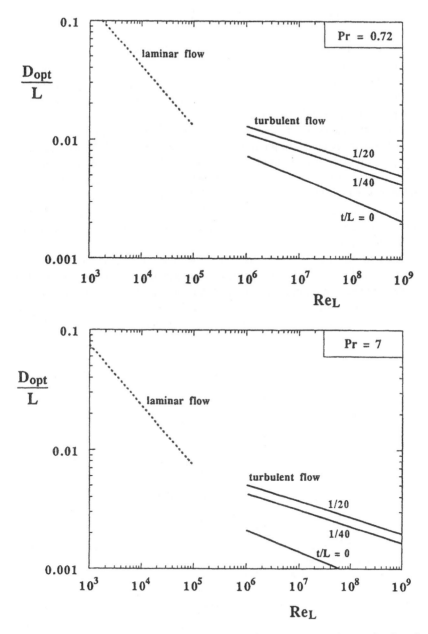

FIGURE 11. The optimal spacing for laminar and turbulent forced convection as a function of the free-stream Reynolds number $Re_L = U_0 L/\nu$ (Bejan and Morega, 1994).

$$\left[\frac{q'L}{kH(T_{max}-T_\infty)}\right]_{max} \leq 0.41\,Pr^{51/99}\left(1+\frac{t}{D_{opt}}\right)^{-67/99}Re_L^{94/99} \qquad (44)$$

The t/D_{opt} ratio appearing on the right side is given by Equation 41. The resulting maximum thermal conductance estimate has been plotted in Figure 12 next to the corresponding curves for laminar flow. The analogy between the use of Re_L on the abscissa (Figures 11 and 12) and the use of the pressure drop number (Figures 9 and 10) is worth noting.

The optimization of the internal geometry of a stack immersed in a free stream was investigated in great detail by Morega et al. [1995]. As shown in Figure 13, it was assumed in the beginning that the plates in the stack are not equidistant. The first objective was to determine the optimal spacing for each individual pair of plates, i.e., the optimal distribution of spacings across the stack. This problem was solved numerically by calculating the temperature and flow fields in a large number of configurations. The plate surfaces were modeled as uniform-flux.

The optimization procedure for a stack with n = 4 nonequidistant plates is illustrated in Figure 14. The design has only one degree of freedom represented by the half-spacing d_1, or the position of the internal plate. The sum of the two spacings is fixed, $d_1 + d_2 = (L/2) - 2t$. For each flow (e.g., $Re_L = 200$ in Figure 14), the temperature distributions over all the surfaces were calculated. There

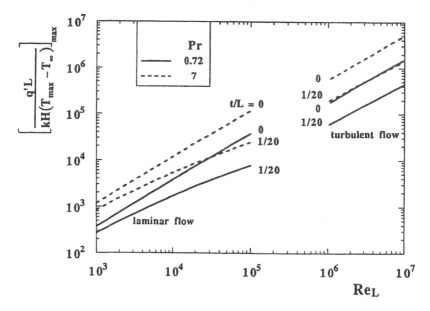

FIGURE 12. The maximum overall thermal conductance for laminar and turbulent forced convection as a function of the free-stream Reynolds number $Re_L = U_0L/v$ (Bejan and Morega, 1994).

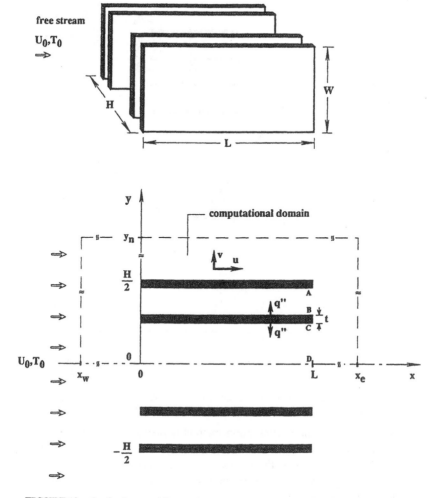

FIGURE 13. Stack of nonequidistant plates cooled by a free stream (Morega et al., 1995).

are four such distributions, θ_1, θ_2, θ_3, and θ_4, which are nondimensionlized as in Equation 46. The location and value of the maximum temperature was identified. The four temperature maxima are plotted in dimensionless terms (θ_{1max}, ..., θ_{4max}) in Figure 14. These temperature maxima occur close to the trailing edge of each surface.

The abscissa of Figure 14 accounts for changes in the relative position of the inner plate. It is clear that this position can be selected such that the peak temperature of the entire stack is minimized. Of interest then is the minimization of the largest of the four temperature maxima, i.e., the minimization of the dimensionless hot-spot temperature,

$$\theta_{hot} = \max \left(\theta_{1,max}, \theta_{2,max}, \theta_{3,max}, \theta_{4,max} \right) \qquad (45)$$

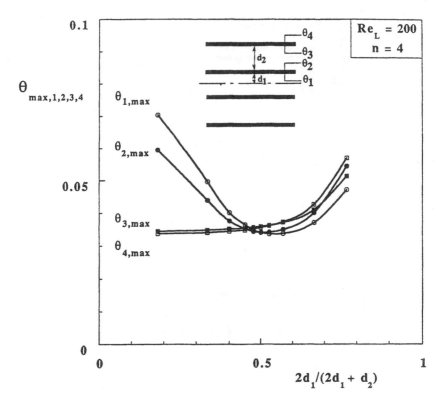

FIGURE 14. Stack immersed in a free stream: the effect of the position of the inner plates on the maximum temperatures of the plate surfaces ($n = 4$, $Re_L = 200$) (Morega et al., 1995).

This second step of the optimization method is presented in Figure 15. The symbols that are superimposed on the curve $Re_L = 200$ indicate which of the four temperature maxima is the largest, i.e., the surface on which the hot spot is located. The hot spot jumps from one surface to another as the position of the inner plate changes.

Next to the $Re_L = 200$ data derived from Figure 14, Figure 15 shows the corresponding results developed for $Re_L = 100$ and $Re_L = 400$. The effect of the inner plate position is clear: the hot-spot temperature θ_{hot} is always the lowest when $2d_1/(2d_1 + d_2)$ is close to 0.5, i.e., when the plates are positioned equidistantly. The minimum exhibited by θ_{hot} is sufficiently flat in the vicinity of $2d_1/(2d_1 + d_2) \cong 0.5$ such that we may conclude with confidence that the optimal design for $n = 4$ and $Re_L = 100$–400 is the one in which the plates are positioned equidistantly. The θ_{hot} minimum becomes flatter as Re_L increases; this means that the relative positioning of the boards becomes less critical as the Reynolds number increases.

These conclusions were reinforced by repeating these calculations for stacks with six plates. The geometry-induced changes in the hot-spot temperature become less pronounced when n increases. This means that the fine-tuning of the

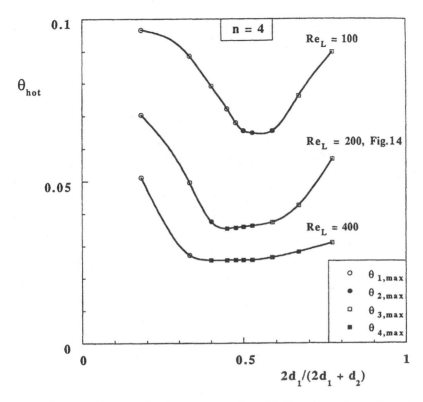

FIGURE 15. Stack immersed in a free stream: the effect of the Reynolds number and the positions of the inner plates on the hot spot temperature of the entire stack (n=4) (Morega et al., 1995).

position of each plate relative to its two neighbors loses its importance as n increases. Since the best designs for n = 4 and n = 6 are the ones in which the plates are spaced equidistantly, it is safe to generalize and to recommend the equidistant spacing as an optimal design feature for stacks with more than six plates.

The second design aspect investigated numerically in Morega, et al. [1995] was the optimal number of plates arranged equidistantly in a stack. The total heat transfer rate from the stack of fixed volume $H \times L \times W$ was independent of the number of boards $q' = \dot{Q}/W$. The hot spot temperature

$$\theta_{hot} = \frac{T_{hot} - T_0}{q'/k} \qquad (46)$$

was calculated for each number of boards (n) and flow regime (Re_L). The results are presented in Figure 16 for the range $2 \leq n \leq 8$ and $100 \leq Re_L \leq 1,000$. It is clear that there exists an optimal number of boards and that knowing this number accurately makes a difference in the effort to maximize the overall thermal conductance (the inverse of θ_{hot}). The $n_{opt}(Re_L)$ values identified in Figure 16 are recorded in Table 1.

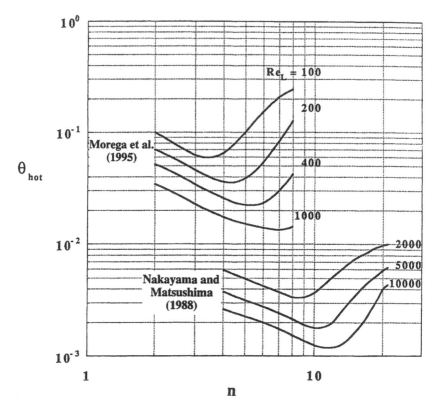

FIGURE 16. Stack immersed in a free stream: the effect of the number of equidistant plates on the hot spot temperature (Morega et al., 1995).

A theoretical n_{opt} formula based on the method of intersecting the asymptotes is [Morega et al., 1995]

$$n_{opt} \cong \frac{0.26 \frac{H}{L} Pr^{1/4} Re_L^{1/2}}{1 + 0.26 \frac{t}{L} Pr^{1/4} Re_L^{1/2}} \qquad (Pr \geq 0.7, n \gg 1) \qquad (47)$$

The n_{opt} values calculated based on Equation 47 have been added to Table 1. The agreement between the rounded (integer) values of these order-of-magnitude estimates and the numerical data furnished by Figure 16 is good, even though the number of boards is small, i.e., outside the range of Equation 47. The relative agreement improves as Re_L increases. This trend is supported further by the independent n_{opt} data collected from [Nakayama et al., 1988] and [Matsushima et al., 1992].

The corresponding minimum hot-spot temperatures were correlated with the theoretical (scaling correct) formula derived in [Morega et al., 1995]

$$\theta_{hot,min} = 1.85 \frac{L}{H} \frac{1 + 3.27(t/L)\,Re_L^{1/2}}{Pr^{1/2}\,Re_L} \qquad (Pr \gtrsim 0.7, n \gg 1) \qquad (48)$$

The Re_L range of this correlation was extended from 100 to 10,000 by also using the data of Nakayama et al. [1988] and Matsushima et al. [1992]. Note that the dimensionless minimum hot-spot temperature $\theta_{hot,min}$ is the inverse of the maximum stack thermal conductance, cf. Equation 46.

When the number of plates is greater than n_{opt}, the flow in each channel approaches the fully developed regime, and the stack as a block can be modeled as a saturated porous medium with zero permeability in the transversal direction [Nield and Bejan, 1992]. It was shown that when the stack porous medium model is combined with the usual pure-fluid model of the external flow the computations are faster and permit a study of the effect of heat conduction through the plate material [Morega et al., 1995].

BUNDLE OF CYLINDERS COOLED BY FORCED CONVECTION

The optimization of the internal geometry of a fixed volume cooled by forced convection can be pursued in other configurations, i.e., in applications where the heating elements are not shaped as parallel plates. An important class is shown in Figure 17; parallel cylinders (isothermal in the W direction) are a good model for pin-fin heat sinks with near 100% fin efficiencies.

The optimal spacing S for cylinders arranged in an equilateral triangular array has been determined based on the method of intersecting the asymptotes [Bejan, 1995a,b]. The asymptotes were determined by using the large volume of empirical data accumulated in the literature for single cylinders (large-S limit) and arrays with many rows (small-S limit) [Zukauskas, 1987]. It was assumed that the cylinder diameter (D) and the pressure drop across the bundle (Δp) are specified, and that each cylinder has the same heat transfer rate. In the range $10^4 \leq \tilde{P} \leq 10^8$, $25 \leq H/D \leq 200$ and $0.72 \leq Pr \leq 50$, the optimal spacing is correlated within 5.6% by the expression

$$\frac{S_{opt}}{D} = 1.59 \frac{(H/D)^{0.52}}{\tilde{P}^{0.13}\,Pr^{0.24}} \qquad (49)$$

where $\tilde{P}$ is a pressure drop number based on D

$$\tilde{P} = \frac{\Delta p\,D^2}{\mu v} \qquad (50)$$

The hot spot of the bundle occurs on the cylinders that occupy the last row. The minimum hot-spot temperature, which corresponds to the optimal spacing of Equation 49, is correlated within 16% by the expression

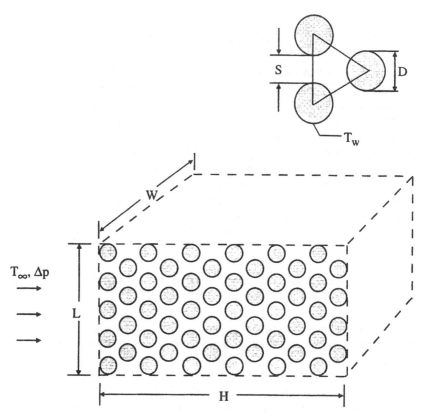

FIGURE 17. Fixed volume with bundle of parallel cylinders perpendicular to a free stream (Bejan, 1995a,b).

$$\frac{T_{max} - T_{\infty}}{\dot{Q}D/(kLW)} \cong \frac{3.33}{\tilde{P}^{0.45} \, Pr^{0.64}} \tag{51}$$

Both Equations 49 and 51 are recommended by the analytical solution obtained by intersecting the asymptotes. Noteworthy is that the denominator on the right side of Equation 51 is approximated well by $\Pi_D^{1/2}$ where the pressure drop number is based on D

$$\Pi_D = \frac{\Delta p D^2}{\mu \alpha} \tag{52}$$

This allows us to rewrite Equation 51 as the maximum power density that can be installed on cylinders in the fixed volume $H \times L \times W$

$$\frac{\dot{Q}}{HLW} \sim 0.3 \, \Pi_D^{1/2} \frac{k}{HD} (T_{max} - T_{\infty}) \tag{53}$$

It is interesting that since Π_D is proportional to D^2 the maximum power density is independent of the cylinder diameter. It is important in thermal design to select S in accordance with Equation 49.

For applications where the free stream velocity U_0 is specified (instead of Δp), Equations 49 and 51 may be transformed by noting that, approximately $\Delta p \sim (1/2) \rho U_0^2$. The results are [Bejan, 1995b]

$$\frac{S_{opt}}{D} \cong 1.7 \frac{(H/D)^{0.52}}{Re_D^{0.26} Pr^{0.24}} \tag{54}$$

$$\frac{T_{max} - T_\infty}{\dot{Q}D/(kLW)} \cong \frac{4.5}{Re_D^{0.9} Pr^{0.64}} \tag{55}$$

where $Re_D = U_0 D/\nu$, or $Re_D \sim (2\tilde{P})^{1/2}$. Equations 54 and 55 cover the range $140 < Re_D < 14,000$, $25 < H/D < 200$, and $0.72 < Pr < 50$.

HEAT GENERATING PLATE COOLED INSIDE A PARALLEL PLATE CHANNEL

In certain types of electronic packages, the single-phase coolant flows through a set of two-dimensional parallel channels formed by a row of printed circuit boards plugged into a mother board. Each board may be surrounded by a metal or metal-coated plastic case whose function is to shield the electronic circuitry from external electromagnetic noise. It is important to know the optimal geometry of each cassette (i.e., the board and its parallel-plate casing) so that the board operating temperature is minimum [Bejan et al., 1993]. To optimize the geometry of the cassette means to find not only the optimal position for the board inside the channel but also the optimal slenderness of the cassette itself (i.e., the spacing of the channel in which the heat generating board is encased).

BOARD WITH LARGE THERMAL CONDUCTANCE IN THE TRANSVERSAL DIRECTION

Consider the problem of cooling in the most effective way a board of length L by positioning it in a stream of coolant that flows through an insulated parallel-plate channel of the same length. The channel spacing D is fixed. The geometry sketched in Figure 18 is two-dimensional, as the board and the channel are sufficiently wide (width W) in the direction perpendicular to the figure, $W > L$.

We assume that the pressure difference across the channel is fixed, Δp, because the flow is driven by a fan with diameter considerably greater than the channel spacing D. In an actual application, the fan would blow air through a stack of ten or more cassettes of profile $L \times D$. One such cassette is presented in Figure 18.

The total rate of heat transfer $\dot{Q}$ from the heated plate to the fluid, through both sides of the plate, is fixed by the electric circuit design. The plate

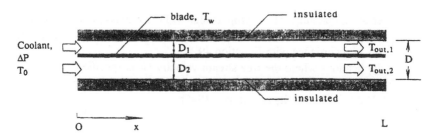

FIGURE 18. Heat generating board cooled by a stream in an insulated parallel-plate channel (Bejan et al., 1993).

thickness is negligible with respect to the channel spacing D. The only degree of freedom in choosing the best cooling arrangement is the position of the heated plate inside the channel. This position is pinpointed by the subchannel spacings above and below the heated plate, D_1 and D_2, such that $D_1 + D_2 = D$. To illustrate the solution method in the simplest possible way, we first assume that:

1. The heated plate is isothermal at T_w.
2. The flow is fully developed and laminar on both sides of the plate.
3. The NTU on either side of the plate is sufficiently greater than 1 so that $(T_{out,1,2} - T_0)/(T_w - T_0) \cong 1$.
4. The surfaces of the plate and the channel walls are smooth.

As shown in Figure 18, $T_{out,1}$ and $T_{out,2}$ are the outlet bulk temperatures above and below the heated plate. The objective is to determine the best configuration (D_1/D) so that the thermal conductance $\dot{Q}/(T_w - T_0)$ is the greatest.

If we label $\dot{Q}_1$ and $\dot{Q}_2$ the heat transfer rates through the upper side and the lower side of the heated plate, then the approximation (1) permits us to write

$$\dot{Q}_1 = \dot{m}_1 c_p(T_w - T_0), \quad \dot{Q}_2 = \dot{m}_2 c_p(T_w - T_0) \tag{56}$$

where, for fully developed laminar flow, [e.g., Bejan, 1993]

$$\dot{m}_1 = \frac{\rho W}{12\mu} \frac{\Delta p}{L} D_1^3, \quad \dot{m}_2 = \frac{\rho W}{12\mu} \frac{\Delta p}{L} D_2^3 \tag{57}$$

Next, we write y and $(1 - y)$ for the dimensionless spacings of the upper and lower subchannels

$$D_1 = y\,D, \quad D_2 = (1 - y)\,D \tag{58}$$

and calculate the total heat transfer rate $\dot{Q} = \dot{Q}_1 + \dot{Q}_2$

$$\frac{\dot{Q}}{T_w - T_0} \frac{12\,\mu L}{\rho c_p\,WD^3\Delta p} = y^3 + (1 - y)^3 \tag{59}$$

Figure 19 shows that the highest value of the y function on the right-hand side is 1 and that it occurs when y = 0 or 1. The minimum value (1/4) occurs when y = 1/2. We reach the somewhat unexpected conclusion that, from a cooling standpoint, the centerplane (y = 1/2) is the worst position that the heated board can have. The best arrangement is the one where the board is attached to one of the insulated walls of the channel, even though in that case the entire heat transfer rate $\dot{Q}$ must leave the plate through only one of its side surfaces. When the board is attached to one of the walls, the thermal conductance $\dot{Q}/(T_w - T_0)$ is four times greater than when the board is positioned in the center of the channel.

The conclusion that the worst cooling position is y = 1/2 remains valid even when some of the simplifying assumptions (1) through (4) are relaxed. For example, let us discard (2) and (4) together and assume instead that the flow is turbulent and fully developed and that the board surfaces are very rough. This is a good model for cassettes with L/D ratios much greater than 10, so that the

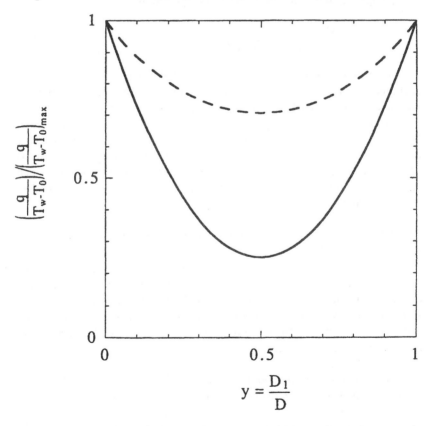

FIGURE 19. The effect of the board position on the overall thermal conductance of the cassette of Figure 18: _____ laminar, fully developed flow, -------- turbulent, fully rough, fully developed flow (Bejan et al., 1993).

entrance region is relatively small, and with boards densely covered with chips and circuitry that rise as large-scale three-dimensional asperities above the surface. Under these circumstances (i.e., in the fully rough limit of turbulent duct flow) the friction factor is practically independent of the Reynolds number, and Equation 57 is replaced by

$$\dot{m}_1 = \left(\frac{\rho\,\Delta p}{f_1\,L}\right)^{1/2} WD_1^{3/2}, \quad \dot{m}_2 = \left(\frac{\rho\,\Delta p}{f_2\,L}\right)^{1/2} WD_2^{3/2} \tag{60}$$

The constant friction factors f_1 and f_2 depend on the dimensions of the roughness elements (assumed the same for both board surfaces) and on the respective subchannel spacings (D_1, D_2). When the board is placed in the stream, i.e., at y values comparable with 1/2, the spacings D_1 and D_2 are also comparable and, as a first approximation, f_1 and f_2 may be taken as equal to the same constant f. This is a conservative approximation to which we shall return in the next paragraph. In the end, Equation 59 is replaced by

$$\frac{\dot{Q}}{T_w - T_0}\left(\frac{f\,L}{\rho\,\Delta p}\right)^{1/2} \frac{1}{c_p WD^{3/2}} = y^{3/2} + (1-y)^{3/2} \tag{61}$$

This result shows that the overall thermal conductance $\dot{Q}/(T_w - T_0)$ is once again minimum if the board is placed in the middle of the channel. The right-hand side of Equation 61 is plotted as a dashed curve in Figure 19 and is valid in the vicinity of y = 1/2. The thermal conductance minimum is not as sharp as for fully developed laminar flow, suggesting that the optimal positioning of the board is not as critical in the fully rough limit. If one is to repeat the analysis and take into account the difference between f_1 and f_2 as the board is positioned close to one of the walls, one would obtain a curve that falls under the dashed curve in Figure 19.

BOARD WITH FINITE THERMAL CONDUCTANCE IN THE TRANSVERSAL DIRECTION

Consider now the more realistic model in which the board of Figure 18 (the substrate of an electronic circuit board) has a finite thermal conductivity k_w and thickness t. The thickness continues to be negligible with respect to D. The two surfaces of the board are loaded equally and uniformly with electronics: the constant heat generation rate per unit board surface is q″. It is important to note, however, that the heat fluxes removed by the two streams are generally not equal because of the conduction heat transfer across the board.

The temperatures of the two board surfaces (T_1, T_2) increase in the downstream direction, and reach their highest levels at the trailing edge, x = L. The objective is to minimize the larger of these two trailing-edge temperatures by choosing the optimal board position y. Space limitations permit us to show only the final results, which are displayed in Figures 20 and 21 in terms of the transversal thermal conductance number

$$B = 12\frac{k_w}{k}\frac{\mu\alpha L^2}{\Delta p\,D^3 t} \tag{62}$$

and the dimensionless temperature ceiling

$$\theta = (T_{max} - T_0)\frac{\rho c_p\,\Delta p\,D^3}{12\mu\,L^2 q''} \tag{63}$$

where T_{max} is the temperature of the hot spot, i.e., the larger of the trailing-edge temperatures of the two board surfaces.

Figure 20 shows that the best board position Y_{min} depends on B, i.e., on the degree to which the board substrate is a good thermal conductor:

1. When B is of the order of 1 or larger, the lowest ceiling temperatures are registered at $y_{min} = 0$ and $y_{min} = 1$, i.e., when the board is positioned close to one of the insulated walls of the channel. The worst position is in the middle of the channel, $y_{max} = 1/2$. These conclusions agree completely with what we learned based on the isothermal board model (i.e., $B \to \infty$).
2. When the board is a poor thermal conductor such that B is smaller than the order of 1, the best position for the board is along the midplane of the D × L channel. The worst position, y_{max}, approaches 0.8 and 0.2 as B decreases.

The transition from conducting boards (1) to poorly conducting boards (2) occurs when B drops below 0.166. It is fascinating that the best location for poorly conducting boards, $y_{min} = 1/2$, happens to be exactly the same as the worst location for highly conducting boards. This observation stresses the crucial importance of the dimensionless number B. This number must be calculated early in order to determine the problem type, (1) or (2).

The lowest trailing-edge temperature ceiling that corresponds to the best location y_{min} is presented as θ_{min} versus B in Figure 21. The same figure shows the uppermost trailing-edge temperature that corresponds to the worst position y_{max}, namely θ_{max}. The lowest temperature ceiling (θ_{min}) is considerably smaller than the highest temperature ceiling (θ_{max}), regardless of the B value. This shows the importance of knowing not only the best design (y_{min}) but also the worst design (y_{max}).

STACKS OF PLATES SHIELDED BY POROUS SCREENS

Microelectronics packaging is subject to several conflicting requirements such as electromagnetic compatibility, acoustic limits, and adequate cooling. Electronic systems and components are often enclosed completely inside conducting cases to minimize radio frequency interference or electromagnetic interference or for protection against airborne particles. These enclosures have the added benefit that they reduce the sound pressure level associated with

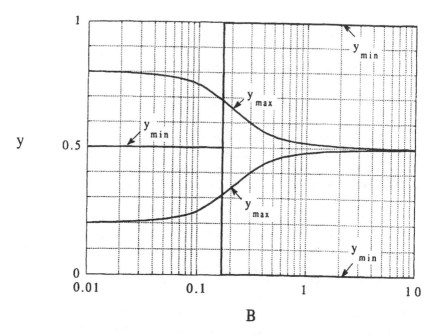

FIGURE 20. The best position (y_{min}) and the worst position (y_{max}) of a heat generating board with finite transversal thermal conductance (Bejan et al., 1993).

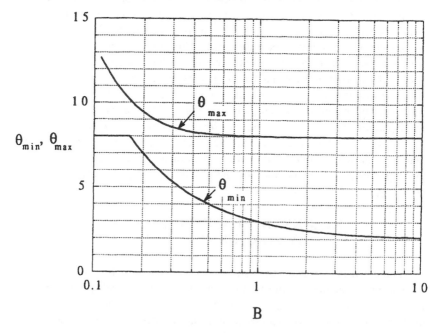

FIGURE 21. The lowest temperature ceiling (θ_{min}) and the highest temperature ceiling (θ_{max}) that correspond to the best location (y_{min}) and the worst location (y_{max}) (Bejan et al., 1993).

acoustic noise generation. There are numerous applications where the thermal design rules out the use of a complete enclosure around the electronic package. Openings must be provided in the enclosure because the cooling requirement of an enclosed packages is greater, and the need to ventilate the electronic components becomes critical. This conflict between the need to enclose and the need to cool poses a significant design challenge as the power density of electronics is increased.

The importance of the optimal selection of openings for air cooling has been recognized in the design of stacks of parallel printed circuit boards. The optimal thermal performance of a stack of parallel boards surrounded by perforated screens formed the subject of a recent study [Bejan et al., 1995]. The study was structured as a sequence of four distinct problems according to the cooling mode (forced convection versus natural convection) and whether the board-to-board spacing is fixed or constitutes a degree of freedom in the design. In every case a relation was established between the characteristics of the perforated screens (e.g., porosity) and the overall thermal conductance between the stack and the coolant. In what follows we illustrate two examples.

FORCED CONVECTION

Consider the L × H stack of heat generating parallel plates (e.g., printed circuit boards) shown in Figure 22. It is assumed that the stack dimension perpendicular to the figure is sufficiently greater than the plate-to-plate spacing D such that the flow through each channel is two dimensional. Two electromagnetic screens (e.g., perforated plates) are placed immediately upstream and downstream of the stack. The coolant (e.g., air) is forced to flow through the entire assembly by the pressure difference Δp, which is fixed. The flow through the spaces between the parallel plates is laminar. The screen is described by the porosity φ and contraction pressure loss coefficient K_0, for which there are sufficient empirical data in the literature (see also Bejan et al., [1995]).

The optimal spacing was determined by using the method of intersecting the asymptotes. The solution is presented in Figure 23 in the dimensionless notation

$$\delta = \frac{D_{opt}}{L} \prod{}^{1/4}, \quad \prod = \frac{\Delta p \cdot L^2}{\mu \alpha}, \quad x = \frac{K_c}{\varphi^2} Pr^{-5/3} \tag{64}$$

The solution accounts for the effect of screen porosity (through φ and K_c), the effect of screen geometry (through K_c: perforated versus woven screen, sharp versus rounded edges), and the effect of fluid type (through Pr). We see that when the screens are absent ($x = 0$) the optimal spacing parameter approaches $\delta = 2.73$, i.e., Equation 22. In the opposite limit ($x \gg 1$), the optimal spacing approaches $\delta = 1.32\, x^{1/4}$. It is worth noting that in both x limits the δ (x) relation is independent of Pr; this feature is the result of including Pr in the definition of x, Equation 64. The Prandtl number has the peculiar effect that it shifts the minimum of the δ (x) curve, however, this is a minor effect in the Pr range 0.72–7.

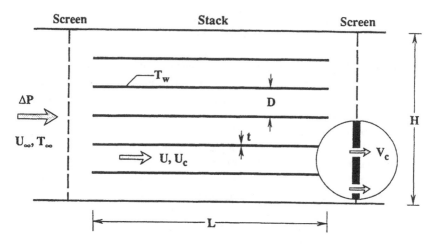

FIGURE 22. Forced convection cooling of a stack of parallel plates with inlet and outlet screens (Bejan et al., 1995).

Along the top of Figure 23 we plotted the porosity that corresponds to the abscissa parameter x when the coolant is air (Pr = 0.72) and the screen is a plate with sharp-edge perforations. For the function $K_c(\varphi)$ we used the values tabulated in [Blevins, 1992]. The important conclusion made visible in Figure 23 is that the optimal spacing increases when the screen becomes an increasingly more significant flow obstruction. The upper abscissa shows that the effect of the screens on the selection of D begins to be felt when the porosity φ drops below approximately 0.7. When screens with porosities greater than 0.7 are used, the optimal plate-to-plate spacing is the same as when the screens are absent.

NATURAL CONVECTION

We now turn our attention to the analogous question for designs where the parallel plates are vertical and the cooling is by natural convection. As shown in Figure 24, most of the modeling features described in connection with Figure 22 are repeated. The assembly is immersed in a quiescent fluid of temperature T_∞. The number of parallel plates, n = L/D, is assumed considerably greater than 1.

The intersection of the small-D and large-D asymptotes produced the optimal spacing reported in Figure 25, where

$$\delta_* = \frac{D_{opt}}{H} Ra_H^{1/4}, \quad x_* = \frac{K_c}{\varphi^2 Pr}, \quad Ra_H = \frac{g\beta H^3 (T_w - T_\infty)}{\alpha\nu} \tag{65}$$

The optimal spacing decreases as the abscissa parameter increases, i.e., as the screens pose an increasing flow resistance. The decrease in δ_* versus x_* is quite interesting because it runs against the trend exhibited by the corresponding result for forced convection, Figure 23. Interesting also are the similarities between Figures 25 and 23, specifically, the near-same abscissa parameters and the fact that the effect of the screens becomes important when the abscissa pa-

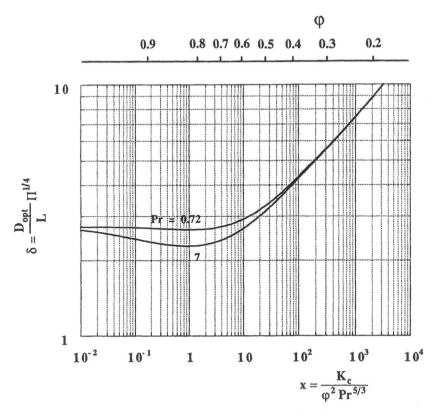

FIGURE 23. The optimal plate-to-plate spacing as a function of the screen characteristics when the stack is cooled by forced convection (high screen Reynolds numbers). The upper φ scale refers to air cooling (Pr = 0.72) and screens made out of plates with sharp-edged perforations (Bejan et al., 1995).

rameter exceeds a critical order of magnitude. In Figure 25 the knee of the δ_* (x_*) curve is located in the vicinity of $x_* \sim 1$, which in the case of air cooling and plate screens with sharp-edged perforations means that the screen effect becomes important when the porosity falls below approximately 0.8.

Several other optimization problems involving stacks with inlet and outlet screens in forced or natural convection can be found in Bejan et al. [1995].

PLATE FINS WITH VARIABLE
THICKNESS AND HEIGHT

The hot-spot temperature of a module with finned air heat sink can be reduced by allowing the fin thickness and height to vary in the flow direction x, while the amount of fin material is fixed [Morega and Bejan, 1994b]. This geometry is illustrated in Figure 26. Two models were used. In a two-dimensional model, the fin conduction was uncoupled from the external convection by as-

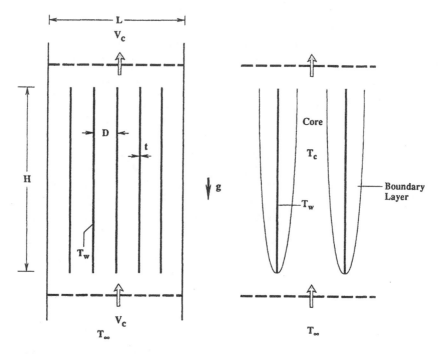

FIGURE 24. Natural convection cooling of a stack of parallel plates with inlet and outlet screens. Right side: distinct boundary layers and core temperature in the large-D limit (Bejan et al., 1995).

suming that the heat transfer coefficient varies as $x^{-1/2}$, in accordance with boundary layer theory. In the second model the three-dimensional problem of conjugate fin conduction and external convection was solved numerically. The main conclusions are as follows:

1. The maximum temperature on the base of a fin with constant height is reduced by approximately 15% if the plate fin is sharpened like a dull knife such that its thickness increases as $x^{0.42}$ in the downstream direction.

2. The hot-spot temperature at the base of a fin with constant thickness is reduced by approximately 30% if the crest is inclined to face the flow, with nearly zero height at the leading edge. The forward inclination of the crest is the result of having assumed that the base of the fin is heated with uniform flux. It can be shown that when the top of the module is conductive enough the fin base is isothermal and the optimal crest inclination has a negative slope, i.e., the crest looks downstream.

3. In addition to lowering the hot spot temperature, each of the design changes (1) and (2) leads to a considerably more uniform temperature distribution on the module surface on which the finned heat sink is installed.

4. It is conceivable that an even greater reduction in hot spot temperature can be achieved by implementing the design features (1) and (2) simul-

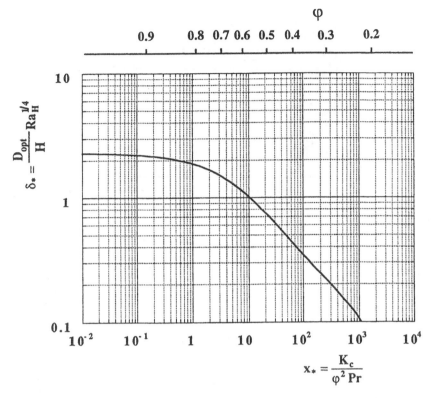

FIGURE 25. The optimal plate-to-plate spacing as a function of the screen characteristics when the stack is cooled by natural convection (high screen Reynolds numbers). The upper φ scale refers to air cooling (Pr = 0.72) and screens made out of plates with sharp-edged perforations (Bejan et al., 1995).

taneously. The determination of the actual magnitude of this combined reduction would be a good topic for a future three-dimensional numerical study of the conjugate fin-conduction and convection problem (including conduction inside the module) in which the sharpness of the plate and tilting of the crest both vary.

5. A simpler way to proceed would be, first, to design the plate fin with constant thickness and height by using the classical method (e.g., Bejan, [1993], pp. 65–67) and, second, to reshape the thickness and height distributions according to conclusions (1) and (2) above, while keeping the fin volume (weight) fixed. Related air-cooled heat sink geometries are discussed in [Homentovschi et al., 1996; Ledezma and Bejan, 1996].

HEAT SINKS WITH PIN FINS AND PLATE FINS

The optimization of heat sinks with long finned channels and fully developed flow in each channel was performed in Tuckerman and Pease [1981], Knight et al. [1991], and Weisberg et al. [1992]. The work summarized in this

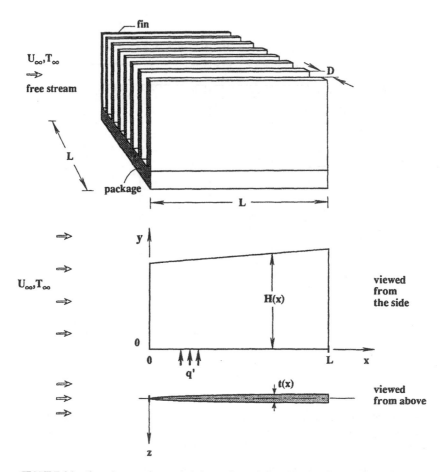

FIGURE 26. Forced convection-cooled electronic package with plate-fin heat sink (top), and two-dimensional conduction model for a single plate fin (bottom) (Morega and Bejan, 1994b).

section refers to three-dimensional arrays of fins, that is, fins that are interrupted (i.e., not long) in the flow direction [Bejan and Morega, 1993]. The optimization was based on an analysis where the space filled by fins and coolant was treated as a porous medium [Bejan, 1990].

PIN FIN ARRAYS

The heat sink with pin fins cooled by forced convection is illustrated in Figure 27. Each pin fin has the diameter d and length L. The area covered by the array is B × X. The generated heat flux is uniform, $q'' = \dot{Q}/BX$. The array has been optimized with respect to two degrees of freedom—first, the pin fin diameter and, second, the array porosity ϕ (or hydraulic diameter D_h). The results of the first maximization of the thermal conductance are shown in Figure 28, where

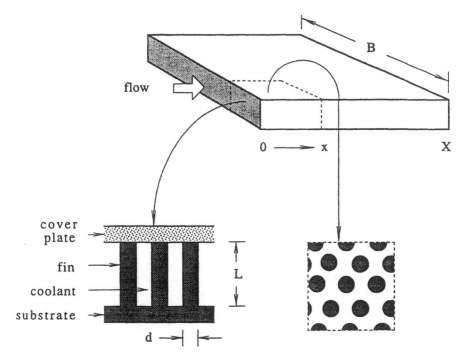

FIGURE 27. Heat sink with array of pin fins cooled by forced convection (Bejan and Morega, 1993).

$$\delta_{opt} = \frac{d_{opt}}{D_h}, \quad b = \frac{L}{D_h}\left(4\,Nu\,\frac{k_f}{k_w}\right)^{1/2} \tag{66}$$

$$G = \frac{\delta^{1/2}}{\delta+1}\tanh\left(\frac{b}{\delta^{1/2}}\right) \tag{67}$$

and Nu is a constant of order 1 [Bejan, 1995a]. In the same solution k_f and k_w are the fluid and wall (fin) thermal conductivities.

The final results for the second maximization of the thermal conductance are shown in Figure 29. This time the hydraulic diameter (or its inverse, b_{opt}) was selected optimally, which is equivalent to selecting the array porosity Φ_{opt}, cf. porous-medium convection theory [Nield and Bejan, 1992]. The resulting twice-minimized thermal resistance $(T_h - T_c)_{min}/(q''L/k_w)$ is also reported. Note that T_h and T_c are the base hot-spot temperature and the coolant inlet temperature. The abscissa of Figure 29 corresponds to the pressure drop number

$$\tilde{P} = 4\frac{Nu}{k_z}\left(\frac{k_f}{k_w}\right)^2 \frac{\Delta p L^4}{\mu\alpha X^2} \tag{68}$$

where k_z is a constant of order 10^2 [Bejan, 1990] and Δp is the pressure difference maintained across the entire heat sink. It is worth noting that $\tilde{P}$ is a multi-

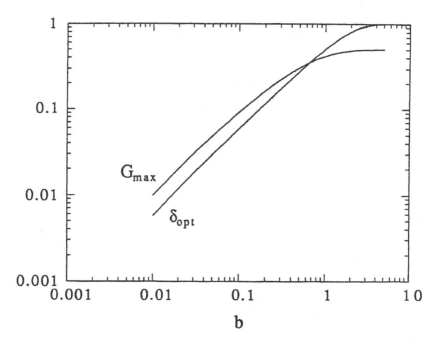

FIGURE 28. Optimal pin fin diameter for an array treated as a porous medium (Bejan and Morega, 1993).

ple of the basic pressure drop number identified in Equation 23. Figures 29 and 28 must be used together with Figure 29 first and $\tilde{P}$ as input. To improve the accuracy of these calculations there is a need for information of heat transfer and pressure drop at low Reynolds numbers [Fowler and Bejan, 1994].

PLATE FIN ARRAYS

The analogous plate-fin geometry is shown in Figure 30. The number of rows is n. The swept length of each plate fin is ξ, the fin thickness is t, and the fin spacing is D. The overall thermal resistance of the heat sink was first minimized with respect to t. The results are shown in Figure 31, where

$$\tau_{opt} = \frac{t_{opt}}{D}, \quad \Phi = 1.1\left(\frac{k_f}{k_w}\right)^{1/2} \frac{L/X}{(D/X)^{1/3}}\left(\frac{n\,\Delta p\,X^2}{\mu\alpha}\right)^{1/6} \tag{69}$$

$$F = \frac{\tau^{1/2}}{(1+\tau)^{5/6}}\tanh\left[\Phi\frac{(1+\tau)^{1/6}}{\tau^{1/2}}\right] \tag{70}$$

The results of minimizing the thermal resistance with respect to the spacing D are shown in Figure 32. The optimal spacing is represented by Φ_{opt}, cf. Equation 69, or by the array porosity ϕ_{opt}. The abscissa parameter is an appropriate pressure drop number

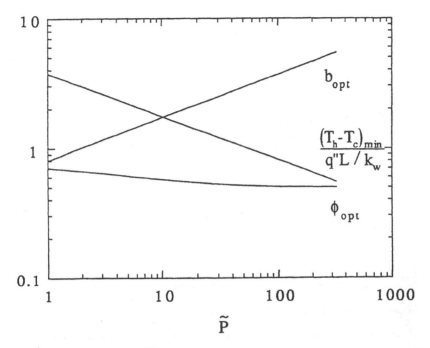

FIGURE 29. Optimal inverse hydraulic diameter (b_{opt}), minimum overall thermal resistance, and corresponding optimal porosity (ϕ_{opt}) (Bejan and Morega, 1993).

$$P_* = \mathrm{Pr}^{1/3} \left(\frac{k_f}{k_w} \right)^2 \frac{\Delta p\, L^4}{\mu \alpha X^2} \qquad (71)$$

that too is a multiple of the pressure drop number defined in Equation 23. The results for optimal plate fin arrays can be used by starting with P_* as input, calculating the optimal spacing (or Φ_{opt}, ϕ_{opt}) from Figure 32 and concluding with the optimal plate-fin thickness from Figure 31. The corresponding minimum thermal resistance is provided by Figure 32.

An important fundamental development that follows from the geometric optimization reviewed in Figures 27–32 is that the flow through a given space filled with complicated objects can be treated as a flow through a *porous medium* at moderate pore Reynolds numbers $UK^{1/2}/\nu$ [Bejan, 1990]. The dimensionless groups used on the ordinate and abscissa in Figure 33 are used routinely in the field of convection in porous media [Nield and Bejan, 1992] where U is the volume averaged velocity. The bundle of four curves corresponds to flow across staggered cylinders (equilateral triangle array) with four (transverse pitch/diameter) ratios: 1.25, 1.5, 2 and 2.5. The data for these curves were taken from Zukauskas [1987], or Bejan [1993, p. 488], and the permeability K was modeled as

$$K = \frac{\phi^3 d^2}{k_z(1 - \phi)^2} \qquad (72)$$

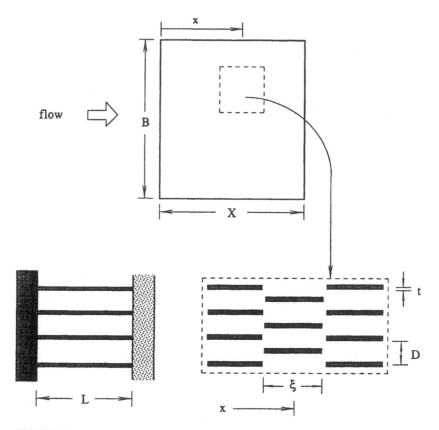

FIGURE 30. Heat sink with array of plate fins cooled by forced convection (Bejan and Morega, 1993).

where $k_z = 100$ and the ϕ is the porosity (void fraction) of the space. The porous medium presentation of the cylinder array Δp leads to a much tighter collapse of the four curves than in the original "heat exchanger handbook" presentation of the same data.

The porous medium presentation of heat exchanger pressure drop information is waiting to be extended to other heat exchanger geometries. In addition to the tight correlation that this method brings, the fact that all the curves must approach the Darcy flow limit

$$\frac{\Delta p\, K^{1/2}}{X \rho U^2} = \left(\frac{UK^{1/2}}{\nu}\right)^{-1} \tag{73}$$

when $UK^{1/2}/\nu < 1$, means that the Δp data can be extended (extrapolated) with confidence into the low Reynolds number range where experimental data are scarce [Fowler and Bejan, 1994]. Furthermore, flows at low Reynolds numbers are becoming more common as the miniaturization of classical heat exchanger configurations continues.

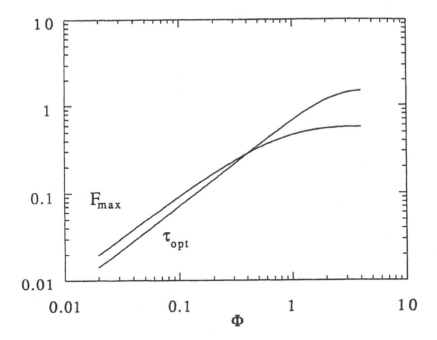

FIGURE 31. Optimal plate-fin thickness for an array treated as a porous medium (Bejan and Morega, 1993).

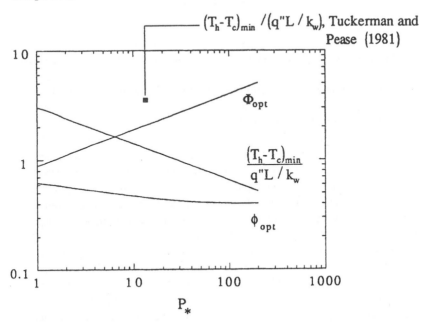

FIGURE 32. Optimal plate-to-plate spacing and porosity, and minimum overall thermal resistance (Bejan and Morega, 1993).

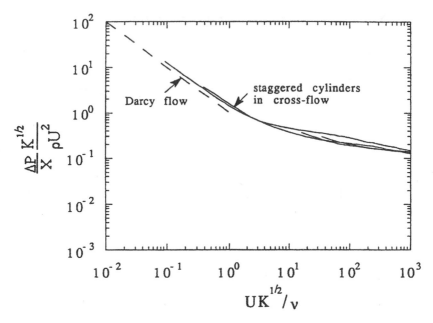

FIGURE 33.   Porous medium representation of the pressure drop data for flow through staggered cylinders (Bejan and Morega, 1993).

CONCLUSION

In this chapter we reviewed a wide selection of heat transfer configurations. In each case we minimized the overall thermal resistance between the coolant and a fixed volume occupied by heat generating components. The heat transfer mechanism varied from natural convection to forced convection. The shape of the components varied from long plates to parallel cylinders and, in the last section, to short plates and short cylinders.

The most important conclusion of this work is that the overall thermal resistance can be reduced geometrically by optimally arranging the heat generating components in the given volume. Quite often, the optimal positions (spacings) can be determined based on simple analyses such as the method of intersecting of asymptotes. The results for optimal spacings and corresponding minimum resistance (or maximum conductance) are now available in compact, scaling correct formulas for key geometries. These results define a newly emerging subfield of heat transfer, namely, the *Geometric Optimization of Cooling Techniques*.

Future studies may focus on other applications, namely fixed spaces with different overall shapes, and components with shapes not considered in this review. In the future work, the body shape correlation method of Refai Ahmed and Yovanovich [1994] may prove very useful. The effect of self-sustained oscillations in the channels [Amon et al., 1992] also deserves to be studied, particularly with respect to determining the optimal board-to-board spacing. It will be interesting to see how the scaling laws and dimensionless groups identified in this chapter can be extended (i.e., generalized) to cover the optimization of future geometries in a field that promises to be very diverse and important.

ACKNOWLEDGMENTS

This work was supported by the IBM Corporation, Research Triangle Park. The guidance received from Dr. Sang W. Lee is gratefully appreciated. The numerical work was supported by a grant received from the North Carolina Supercomputing Center.

REFERENCES

Amon, C.H., Majumdar, D., Herman, C.V., Mayinger, F., Mikic, B.B., and Sekulic, D.P. 1992. Numerical and experimental studies of self-sustaining flows in communicating channels, *Int. J. Heat Mass Transfer*, 35:3115–3129.

Anand, N.K., Kim, S.H., and Fletcher, L.S. 1992. The effect of plate spacing on free convection between heated parallel plates, *J. Heat Transfer*, 114:515–518.

Bar-Cohen, A. and Rohsenow, W.M. 1984. Thermally optimum spacing of vertical, natural convection cooled, parallel plates, *J. Heat Transfer*, 106:116–123.

Bejan, A. 1984. *Convection Heat Transfer*, John Wiley & Sons, New York, prob 11, 157.

Bejan, A. 1990. Theory of heat transfer from a surface covered with hair, *J. Heat Transfer*, 112:662–667.

Bejan, A. 1993. *Heat Transfer*, John Wiley & Sons, New York.

Bejan, A. 1995a. *Convection Heat Transfer*, 2nd ed., Wiley, New York.

Bejan, A. 1995b. The optimal spacing for cylinders in cross-flow forced convection. *J. Heat Transfer*, 117: 767–770.

Bejan, A. and Lee, S.W. 1994. Optimal geometry of convection cooled electronic packages, in *Cooling of Electronic Systems*, S. Kakac, H. Yuncu, and K. Hijikata, Eds., Kluwer Academic Publishers, Dordrecht, The Netherlands, 277–291.

Bejan, A. and Morega, A.M. 1993. Optimal arrays of pin fins and plate fins in laminar forced convection, *J. Heat Transfer*, 115:75–81.

Bejan, A. and Morega, A.M. 1994. The optimal spacing of a stack of plates cooled by turbulent forced convection, *Int. J. Heat Mass Transfer*, 37:1045–1048.

Bejan, A. and Sciubba, E. 1992. The optimal spacing of parallel plates cooled by forced convection, *Int. J. Heat Mass Transfer*, 35:3259–3264.

Bejan, A., Fowler, A.J., and Stanescu, G. 1995a. The optimal spacing between horizontal cylinders in a fixed volume cooled by natural convection, *Int. J. Heat Mass Transfer*, 38:2047–2055.

Bejan, A., Kim, S.J., Morega, A.M., and Lee, S.W. 1995. The cooling of stacks of plates shielded by porous screens, *Int. J. Heat Fluid Flow*, 16:16–24.

Bejan, A., Morega, Al., Lee, S.W., and Kim, S.J. 1993. The cooling of a heat generating board inside a parallel-plate channel, *Int. J. Heat Fluid Flow*, 14:170–176.

Bhattacharjee, S. and Grosshandler, W.L. 1988. The formation of a wall jet near a high temperature wall under microgravity environment, *ASME HTD*, 96:711–716.

Blevins, R.D. 1992. *Applied Fluid Dynamics Handbook*, Krieger, Malabar, FL.

Farhanieh, B., Herman, C., and Sunden, B. 1993. Numerical and experimental analysis of laminar fluid flow and forced convection heat transfer in a grooved duct, *Int. J. Heat Mass Transfer*, 36:1609–1617.

Fowler, A.J. and Bejan, A. 1994. Forced convection in banks of inclined cylinders at low Reynolds numbers, *Int. J. Heat Fluid Flow*, 15:90–99.

Homentovschi, D., Stanescu, G. and Bejan, A. 1996. Cooling of a two demensional space with one or more streams making one or more passes, *Int. J. Heat Fluid Flow*, 17: in press.

Ishizuka, M. 1992. Application of thermo-fluid analysis to the thermal design of air-cooled electronic equipment, in *Transport Phenomena in Heat and Mass Transfer*, J.A. Reizes Ed., Elsevier, Amsterdam, 1067–1088.

Ishizuka, M. 1993. Thermal design approach for electronic equipment by use of personal computers, in *Computers and Computing in Heat Transfer Science and Engineering*, W. Nakayama and K.T. Yang, Eds., CRC Press, Boca Raton, FL, 391–407.

Kim, S.H. and Anand, N.K. 1994a. Laminar developing flow and heat transfer between a series of parallel plates with surface mounted discrete heat sources, *Int. J. Heat Mass Transfer*, 37:2231–2244.

Kim, S.H. and Anand, N.K. 1994b. Turbulent heat transfer between a series of parallel plates with surface-mounted discrete heat sources, *J. Heat Transfer*, 116:577–587.

Knight, R.W., Goodling, J.S., and Hall, D.J. 1991. Optimal thermal design of forced convection heat sinks-analytical, *J. Electron. Packag.*, 113:313–321.

Ledezma, G. and Bejan, A. 1996. Heat sinks with sloped plate fins in natural and forced convection. *Int. J. Heat Mass Transfer*, 39: in press.

Li, W., Kakac, S., Hatay, F.F., and Oskay, R. 1993. Experimental study of unsteady forced convection in a duct with and without arrays of block-like electronic components, *Wärme und Stoffübertragung*, 28:69–79.

Matsushima, H., Yanagida, T. and Kondo, Y. 1992. Algorithm for predicting the thermal resistance of finned LSI packages mounted on a circuit board, *Heat Transfer Japan Res.*, 21(5):504–517.

Mereu, S., Sciubba, E., and Bejan, A. 1993. The optimal cooling of a stack of heat generating boards with fixed pressure drop, flowrate or pumping power, *Int. J. Heat Mass Transfer*, 36:3677–3686.

Moffat, R.J. and Ortega, A. 1988. Direct air cooling of electronic components, in *Advances in Thermal Modeling of Electronic Components and Systems*, Vol. I, A Bar-Cohen and A.D. Kraus Eds., Hemisphere, New York, 129–282.

Morega, A.M. and Bejan, A. 1994a. Optimal spacing of parallel boards with discrete heat sources cooled by laminar forced convection. *Num. Heat Transfer, Part A*, 25:373–392.

Morega, A.M. and Bejan, A. 1994b. Plate fins with variable thickness and height for air-cooled electronic modules, *Int. J. Heat Mass Transfer*, 37 (Suppl. 1):433–445.

Morega, A., Bejan, A., and Lee, S.W. 1995. Free stream cooling of a stack of parallel plates, *Int. J. Heat Mass Transfer*, 38:519–531.

Nakayama, W., Matsushima, H., and Goel, P. 1988. Forced convective heat transfer from arrays of finned packages, in *Cooling Technology for Electronic Equipment*, W. Aung Ed., Hemisphere, New York, 195–210.

Nield, D.A. and Bejan, A. 1992. *Convection in Porous Media*, Springer Verlag, New York.

Paoletti, S., Rispoli, F., and Sciubba, E. 1989. Calculation of exergetic losses in compact heat exchanger passages, *ASME AES*, 10(2):21–29.

Peterson, G.P. and Ortega, A. 1990. Thermal control of electronic equipment and devices, *Adv. Heat Transfer*, 20:181–314.

Petrescu, S. 1994. Comments on the optimal spacing of parallel plates cooled by forced convection, *Int. J. Heat Mass Transfer*, 37:1283.

Refai Ahmed, G. and Yovanovich, M.M. 1994. Approximate solution of forced convection heat transfer from isothermal simple body shapes. AIAA Paper 94-1971, 6th AIAA/ASME Joint Thermophys, Heat Transfer Conf., Colorado Springs, June 20–23.

Tuckerman, D.B. and Pease, R.F.W. 1981. High-performance heat sinking for VLSl, *IEEE Electron Device Lett*, EDL-2:126–129.

Weisberg, A., Bau, H.H., and Zemel, J.N. 1992. Analysis of microchannels for integrated cooling, *Int. J. Heat Mass Transfer*, 35:2465–2474.

Zukauskas, A. 1987. Heat transfer from tubes in crossflow, *Adv. Heat Transfer*, 18:87–159.

Chapter 2

ENTRANCE DESIGN CORRELATIONS
FOR CIRCUIT BOARDS IN FORCED-AIR COOLING

M. Faghri, M. Molki, and Y. Asako

CONTENTS

INTRODUCTION

This chapter describes an experimental study of pressure drop, heat transfer, and wake effect in the entrance region of an array of rectangular blocks in a duct. The main focus of the work is to present correlations for pressure drop, convective heat transfer coefficient, and thermal wake effect. Also, qualitative information will be provided on the nature of the flow field in this region. The correlations presented here are intended to be used by the practitioners to predict the temperature of the electronic components that have a similar geomet-

47

rical layout and especially short entrance lengths. The range of parameters is close to that often encountered in the computer industry.

A search of literature revealed a number of related studies. Sparrow et al. [1982, 1983] reported heat transfer and pressure drop in arrays of rectangular modules with barriers and missing modules. The focus of their work was to study the effect of missing modules and barriers on thermal-hydraulic behavior of rectangular arrays. Their pressure results for no-barrier arrays without missing modules were obtained for one particular geometry around Reynolds number of 6900.

In an experimental effort, Lehmann and Wirtz [1985] studied the effect of streamwise spacing and length on convection from an array of two-dimensional modules. They also performed visualization tests in the periodically fully developed region of duct and obtained some information on the nature of flow field in this region. The pressure drops reported in this reference is limited to Re = 1000, 2000, and 3000. Some pressure results are also reported by Moffat et al. [1985] for rather tall and widely spaced cubical elements.

For sparse arrays of elements, the pressure drop is mainly determined by form drag and thus the pressure coefficient is independent of Reynolds number. In this connection, the pressure drop coefficients reported in some publications are independent of Re (e.g., Moffat et al., [1985]), while the others (e.g., Tai and Lucas, [1985], Souza Mendes and Santos, [1987], and Hollworth and Fuller, [1987]) showed their pressure results in terms of Re.

Another issue of importance in flow through array of rectangular blocks is to determine the flow regime. In a paper by Garimella and Eibeck [1992], the onset of transition from laminar to turbulent is investigated. They concluded that, in a three-dimensional situation, transition is not only a function of flow rate and geometry but also is a function of location in the array. They have further stated that the transition Re (based on channel height) varies from 700 to 1900, and by increasing the streamwise spacing between the elements, the transition occurs at lower Reynolds numbers.

Arvizu and Moffat [1982] presented a superposition method to predict the temperature distribution in a regular array of cubical elements. Another related work on the subject is a paper by Wirtz and Dykshoorn [1984] in which some information is provided about heat transfer in the entry region of a sparse array of elements.

In a series of experimental investigations, Moffat and his group have extensively studied various aspects of thermal-hydraulic behavior of the arrays of electronic components (Moffat et al. [1985], Moffat and Anderson [1988], Anderson and Moffat [1990], and Anderson and Moffat [1991]). The overall objective of their effort has been to develop the techniques and the data bases needed to predict the operating temperature of the blocks. They have also emphasized the very crucial point that the adiabatic heat transfer coefficients, which are reported in the electronics cooling literature, should be used in reference to the adiabatic fluid temperature. Other relevant studies are those reported by

Sridhar et al. [1990], Faghri et al. [1995], Wirtz and Weiming [1991], Kang [1992], and a review article by Peterson and Ortega [1990]. Despite all the useful information and data available in the aforementioned references, a concise and convenient correlation for temperature prediction is not seen in the open literature, especially for short channels encountered in electronic equipment.

There are several novel aspects to the results presented here. The focus of this work is on the entrance pressure drop, heat transfer coefficients, and the associated thermal wake effects. The data are successfully brought together by defining modified parameters, and they are presented by simple correlations. These correlations are subsequently incorporated into a simple algorithm to estimate the operating temperature of a circuit board with random heating. Further, the flow field is visualized by a simple technique, and the results are discussed.

In this study, an array of rectangular blocks (modules) is positioned along the lower wall of a rectangular duct. These blocks, which represent a model for the modular electronic components, are arranged in an in-line fashion. The geometric variables, namely the module dimension (L), the module height (B), the inter-module spacing (S), and the height of the flow passage between the module and the opposite wall of the duct (H), are varied in such a manner that $B/L = 0.5$, $S/L = 0.125, 0.33, 0.5$, and $H/L = 0.125, 0.25, 0.5, 0.75, 1.0$, and 1.5. These dimensions are close to those often encountered in the computer industry. The working fluid is air, and the Reynolds number based on H and the air velocity in the bypass channel (i.e., the channel formed between the top surface of modules and the opposite wall of the duct) ranges from 400 to 15000.

Attention is now turned to a full description of the work. First, a new correlation will be presented for pressure drop, followed by the presentation of correlations for heat transfer and thermal wake effect. The chapter will conclude with a discussion on how the array temperatures can be predicted by a computer program.

A NEW CORRELATION FOR PRESSURE DROP

THE EXPERIMENTAL APPARATUS AND PROCEDURE

The schematic view of the experimental setup is shown in Figure 1. Laboratory air is drawn into the apparatus through a bell-mouth inlet geometry. The air then passes through a flow development section (559 mm), flow straightener, test section, flow redevelopment section (591 mm), venturimeter, and main valve and then leaves the flow circuit through the blower operating in suction mode.

The air flow is adjusted by the main valve. There is also available a bypass valve that allows for a more precise control of the air flow. The flow rate is measured by a precalibrated venturimeter.

The test section and the corresponding upstream and downstream ducts have a rectangular cross section. The width of the cross section is equal to $W = 178$ mm, but its height (H + B in Figure 1B and C), depending on the dimensions of the array, varies from 12.8 to 76.2 mm.

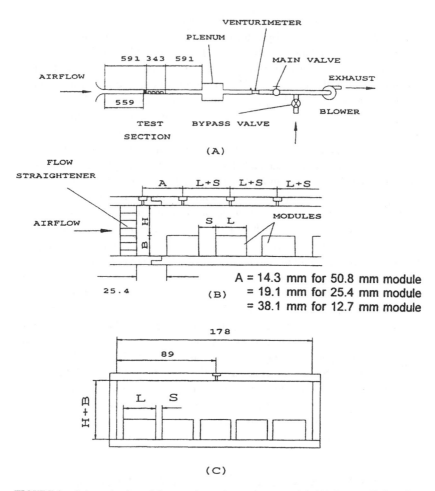

FIGURE 1.   Schematic view of the experimental setup (not to scale). (A) the overall view, (B) side view of the test section, (C) cross-sectional view of the test section. All dimensions are in millimeters.

An in-line array of rectangular modules are deployed along the lower wall of the test section. The dimensions, as noted in Figures 1B and C, are S = 6.4 and 8.4 mm; H = 6.4, 12.7, 19.1, and 25.4 mm; L = 12.7, 25.4, and 50.8 mm; and B = 6.4, 12.7, and 25.4 mm. From the combination of these dimensions nine test sections are fabricated with dimensions shown in Table 1. In all cases B/L is equal to 0.5. The number of rectangular blocks depends on the test section number. For test sections (1 to 3), (4 to 6), and (7 to 9) the number of modules along and across the test section are, respectively, 5 × 3, 8 × 5, and 15 × 9.

The pressure taps are located at the upper wall. As shown in Figure 1B, the first tap is located a distance A upstream of the second tap, and the other taps are equally spaced. The distance A between the first and the second tap de-

TABLE 1
Geometrical Dimensions of the Test Section

Test Section Number	B/L	S/L	H/L	B [mm]	L [mm]	H [mm]	S [mm]
1	0.5	0.125	0.125	25.4	50.8	6.4	6.4
2	0.5	0.125	0.25	25.4	50.8	12.7	6.4
3	0.5	0.125	0.5	25.4	50.8	25.4	6.4
4	0.5	0.33	0.25	12.7	25.4	6.4	8.4
5	0.5	0.33	0.5	12.7	25.4	12.7	8.4
6	0.5	0.33	0.75	12.7	25.4	19.1	8.4
7	0.5	0.5	0.5	6.4	12.7	6.4	6.4
8	0.5	0.5	1	6.4	12.7	12.7	6.4
9	0.5	0.5	1.5	6.4	12.7	19.1	6.4

pends on the test section and is equal to 14.3 (for L = 50.8), 19.1 (for L = 25.4), and 38.1 (for L = 12.7) mm. It should be noted that the location of the first tap was always fixed. However, to position one tap right above the modules it was required to have different A values for different module lengths. In the spanwise direction, the pressure taps are located in the middle of the cross section and just above the central module (Figure 1C). The location of all pressure taps in terms of their distance from the first tap is shown in Table 2. It should be noted that the number of pressure taps for test sections 1 to 3 is more than what is shown in Figure 1B (see Table 2 for exact location of pressure taps).

Prior to the onset of a data run, the proper test section is selected and assembled. Then, to prevent air from leaking into the test section, all suspected joints are sealed by silicon rubber and thoroughly tested for leaks with the aid of soap solution.

Once the test section is properly assembled, the air flow is activated and adjusted for a given flow rate. After a warm-up period, the setup is ready for pressure readings. The pressure signals are transmitted via plastic tubing to a (1 or 10 torr) pressure transducer and an electronic manometer, which are interfaced with an IBM PC. The computer scans the pressure signals for 22 seconds. This arrangement resolves pressure to within 10^{-5} mm Hg. Each experiment is repeated twice and the results are averaged to obtain a data point. A total of 1420 data points are reported here and are used to obtain the pressure drop correlation.

All nine test sections are made entirely of plexiglass so that they can also be used in flow visualization experiments. The flow visualization method used in this work is the so-called oil-lampblack technique.

The results of this investigation are presented in terms of Reynolds number Re with the conventional definition, $Re = VH/v$. In this equation, V is the mean velocity of air in the bypass channel above modules and is related to the volume flow rate as, $V = Q/(WH)$. With this definition, the flow rate between the

TABLE 2
Location of Pressure Taps

Tap Number	Test Sections 1,2,3 X [mm]	Test Sections 4,5,6 X [mm]	Test Sections 7,8,9 X [mm]
1	0	0	0
2	14.3	19.1	38.1
3	28.6	52.9	57.2
4	42.9	86.7	76.3
5	57.2	120.5	95.4
6	71.4	154.3	114.5
7	85.7	188.1	133.6
8	100.0	221.9	152.7
9	114.3	255.7	171.8
10	128.6	289.5	190.9
11	142.9		210
12	157.2		229.1
13	171.5		248.2
14	185.7		267.3
15	200.0		286.4
16	214.3		305.5
17	228.6		324.6
18	242.9		
19	257.2		
20	271.5		

modules at $0 < y < B$ is neglected and V is slightly higher than the actual mean velocity. The pressure drops are nondimensionalized and expressed as

$$K = \frac{P_0 - P_i}{0.5\rho V^2} \qquad (1)$$

where K is the pressure drop coefficient. In the periodically fully developed region, the pressure drops are converted to friction factor defined as

$$f_{2H} = \frac{-\left[\dfrac{dP}{dX}\right] 2H}{0.5\rho V^2} \qquad (2)$$

EXPERIMENTAL UNCERTAINTY FOR PRESSURE DATA

Prior to the onset of experimental runs, the venturimeter was calibrated with the aid of a laminar-flow element. The parameters recorded during the calibration process and the respective measurement uncertainties are listed in Table 3. The precision limits seen in the table are the smallest interval between the scale markings (least count) of the respective instruments. The bias limit for instruments was negligible. The sensitivity coefficients were evaluated using the data reduction FORTRAN program. The results were then combined through the root-sum-square expression to obtain uncertainties (Kline [1985] and Abernethy et al. [1985]).

Using these guidelines, the uncertainty of venturimeter coefficient is found to be ± 0.027 with $C_D = 0.971 \pm 2.8\%$. Other uncertainty values are summarized in Table 4.

FLOW VISUALIZATION

The flow pattern in the entrance region of the array of modules was revealed by the application of oil-lampblack technique. The technique is relatively simple but may not be suitable for any type of flow field. The detailed description of oil-lampblack visualization technique is described by Sparrow et al. [1981].

The first step is to make a suitable mixture of oil and lampblack. It was found that the automatic transmission oil gives the best result. The mixture is subsequently applied to the top surface of the module, which creates a uniform glossy black surface.

When the airflow is initiated, the mixture moves under the action of shear stress exerted by the air, and the streaks that form on the surface reveal the pattern of flow field adjacent to the wall. The development of the surface patterns is visually observed during the visualization run. The experience gained from these visual observations together with the information reflected by the streaks are used to describe the flow field.

TABLE 3
The Precision Limits for the Measured Parameters

Parameter	Precision Limit
Barometric pressure	± 0.1 mm Hg
Venturimeter inlet gage pressure	$\pm 10^{-5}$ mm Hg
Pressure drop across the venturimeter	$\pm 10^{-5}$ mm Hg
Pressure drop across the laminar-flow element	$\pm 10^{-5}$ mm Hg
Inlet air temperature	$\pm 0.6°C$
Laminar-flow element	$\pm 1\%$ of volume flow rate
Duct width, W	± 0.1 mm
Distance between top of modules and the opposite wall, H	± 0.1 mm
Module height, B	± 0.1 mm
Inlet and throat diameters of venturimeter	± 0.1 mm
Pressure gradient, dP/dX	Twice the standard deviation of the slope of P-X plot
test section gage pressure	$\pm 10^{-5}$ mm Hg

TABLE 4
Estimated Typical Uncertainties

Parameter	Uncertainty (%)
Venturimeter coefficient, C_D	2.8
Reynolds number, Re	2.9
Pressure drop coefficient, K	6.1
Friction factor, f_{2H}	15.3

The photographic evidence of the flow field is shown in Figure 2. The photograph shows the top view of the flow pattern near the top surface of the first three modules of the array. These patterns are obtained for the modules with S/L = 0.125 and H/L = 0.125 at Re = 15000. The main flow is from left to right, and the scale below the photograph shows the distance from the leading edge of the first module in centimeters.

Special features of the flow are flow separation at the leading edge of the first module, reattachment of air flow at about 1.8 cm from the leading edge (nearly L/3), and the formation of a recirculating bubble in between. The curved dark line that starts at the lower left corner of the first module and extends to 1.8 cm downstream and subsequently returns to the upper left corner is the line of flow reattachment.

The recirculating bubble is located between the line of reattachment and the leading edge of the first module. In this region, air moves upstream near the wall to meet the incoming separated flow. The dark area extending from the zero of scale to about 0.5 cm downstream (see the scale below the photograph in Figure 2) is simply an area of rather weak shear stress. Beyond the line of reattachment, the air flows downstream toward the other modules of the array.

The patterns clearly show that the flow field near the first module of the array is three dimensional. This is also confirmed by the observations of Chou and Lee [1988] who employed a two-module array in their experiments. Farther downstream the air appears to flow along the direction of main stream and there is no clear evidence of flow separation.

Another message from these photographs is that, at S/L = 0.125 (i.e., the value at which the visualization experiments were performed), there is no clear indication that the flow penetrates the intermodular gaps, as is often observed for larger values of S/L (e.g., see Chou and Lee, 1988). However, the sharp edges of the modules are expected to be conducive to flow separation, and possibly the flow separates behind the modules. This is even more possible at higher Reynolds numbers. If this is true, the pressure results may reflect a lack of dependance on Re, especially at higher Re.

AIRFLOW

→

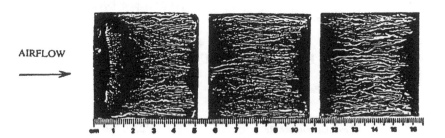

FIGURE 2.   Flow visualization patterns (Re = 15000, L = 50.8 mm, S/L = 0.125, H/L = 0.125).

PRESSURE RESULTS

Pressure distribution along the flow at Re = 15000, S/L = 0.125, and H/L = 0.125 is shown in Figure 3. The lower diagram in this figure is prepared to show the location of pressure taps (marked with X) relative to the modules. As shown, the fist tap is located just upstream of the leading edge of the first module. The ordinate in Figure 3 indicates the air pressure with respect to the atmosphere $P_{atm} - P_i$, so that an increase in the ordinate reflects a drop in pressure.

Examination of the graph indicates that pressure decreases rapidly from the first to the second taps, while there is a slight pressure recovery for the next pair of taps. From this point on the air pressure continuously decreases. The rapid pressure drop between the first and second taps occurs at a point where flow is separated from the wall and the recirculating zone has reduced the effective cross section of the airflow. Beyond this point, the flow expands and the pressure is somewhat recovered.

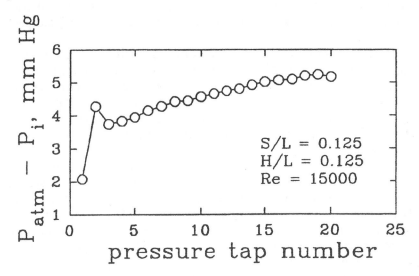

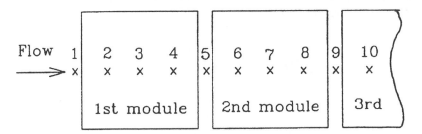

FIGURE 3. Pressure distribution versus tap number at Re = 15000, S/L = 0.125, and H/L = 0.125.

Pressure distributions along the flow direction are shown in Figures 4 to 6. The abscissa X/D_h is the axial distance from the first pressure tap divided by the hydraulic diameter of the duct ($D_h = 2W(H + B)/(W + H + B)$). The ordinate K is the pressure drop coefficient as defined in Equation 1.

Examination of these figures indicates a fairly rapid drop in air pressure as flow enters the array of modules. Further downstream, however, the pressures at similar locations approach a fully developed state and decrease linearly. When the numerical values of pressure data at respective locations were closely examined, it was found that a fully developed state was reached after the 2nd, 3rd, and 4th modules in test sections numbered (1, 2, 3), (4, 5, 6), and (7, 8, 9), respectively.

Another noteworthy feature in these figures is the independence of K from Re when Re is greater than 4000. At higher Re, the possible flow separation behind the modules is intense. In this case, the drag force exerted by the modules on air flow are determined by form drag and the effect of wall friction is negligible. It is known that form drag, and thus K, is nearly independent of Re.

Fully developed pressure drops are expressed in terms of friction factor, f_{2H}, in Figure 7. As stated in the preceding paragraph, there are two mechanisms that govern the pressure drop in the array, namely, the form drag and wall friction. At larger intermodular spacing, the flow is governed by form drag, while at smaller spacing the wall friction is dominant. The data points in Figure 7 indicate that as S/L is increased, the f_{2H} values become less dependent on Re.

Also shown in Figure 7 are the laminar flow results of Asako and Faghri [1988] (the solid line) (at B/L = 0.375 as compared to 0.5 in this investigation), their turbulent results (Asako and Faghri [1991] at B/L = 0.5), and the laminar Hagen-Poiseuille flow in parallel plate channel (the dashed line just above the upper solid line) indicated by 48/Re. There is generally a close agreement between the data, and the minor differences seen in the laminar region is perhaps due to differences in B/L values.

As a first step in obtaining a correlation for the pressure data, we combine the different H/L values with the pressure drop coefficient K in Figure 8. In addition to the experimental data, the figure also shows the best fit through the points. The equation for the solid lines, obtained through a least-squares curve fitting procedure, are

$$K\left[\frac{H}{L}\right]^{-0.0723} = 165.33\frac{\left[\frac{X}{D_h}\right]}{Re} + 0.518 \tag{3}$$

$$K\left[\frac{H}{L}\right]^{-0.393} = 114.21\frac{\left[\frac{X}{D_h}\right]}{Re} + 0.588 \tag{4}$$

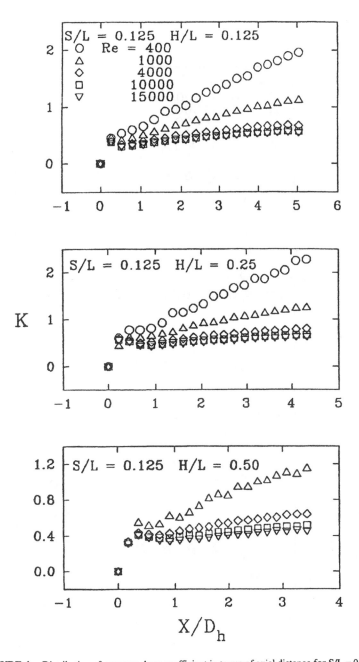

FIGURE 4. Distribution of pressure drop coefficient in terms of axial distance for $S/L = 0.125$.

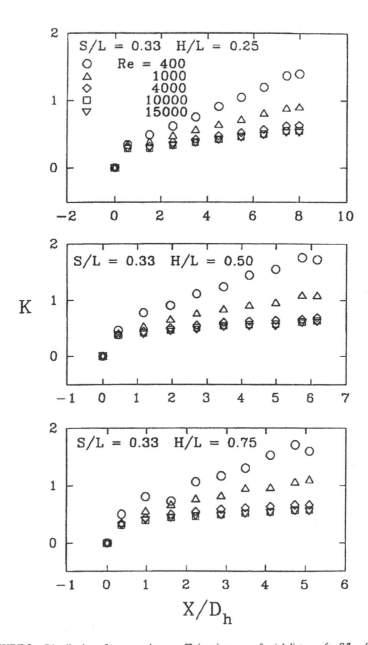

FIGURE 5.   Distribution of pressure drop coefficient in terms of axial distance for S/L = 0.33.

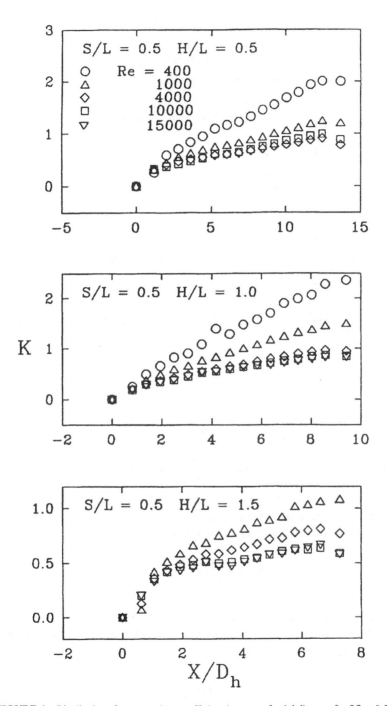

FIGURE 6. Distribution of pressure drop coefficient in terms of axial distance for S/L = 0.5.

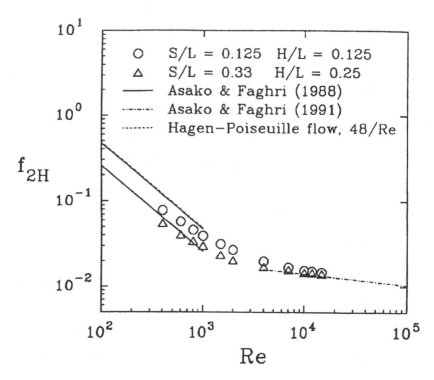

FIGURE 7. Friction factor and comparison with literature.

$$K\left[\frac{H}{L}\right]^{-0.228} = 71.68\frac{\left[\dfrac{X}{D_h}\right]}{Re} + 0.498 \tag{5}$$

The standard deviations of data from the correlations 5 to 7 are, respectively, 0.0677, 0.0760, and 0.1187. With these numbers, the respective correlations are good to within ±17.0%, ±16.9%, and ±30.1%.

Further attempts to correlate all data points and to present them by a single equation requires the definition of modified pressure drop coefficient as

$$K^+ = K\left[\frac{H}{L}\right]^{2.559(\frac{S}{L})^{2.81}} \tag{6}$$

The upper graph in Figure 9 presents a total of 1420 data points. The figure indicates that the data points have been brought together relatively well, and they are represented by the equation

$$K^+ = 87.223\frac{\left[\dfrac{X}{D_h}\right]}{Re} + 0.515 \tag{7}$$

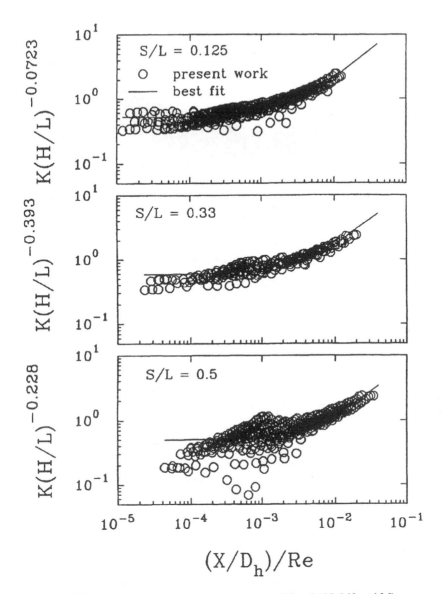

FIGURE 8. The first step in correlating pressure data (S/L = 0.125, 0.33, and 0.5).

The standard deviation of data from correlation is 0.1899 and the correlation is good to within ±54.4%. Further, the standard deviation of coefficients 87.223 and 0.515 are, respectively, $\sigma_1 = 1.493$ and $\sigma_2 = 0.6579 \times 10^{-2}$, corresponding to $\pm 2\sigma_1/87.223 = \pm 3.4\%$ and $\pm 2\sigma_2/0.6579 \times 10^{-2} = \pm 2.6\%$ uncertainty. This equation provides a means for estimating the pressure drop in the entrance re-

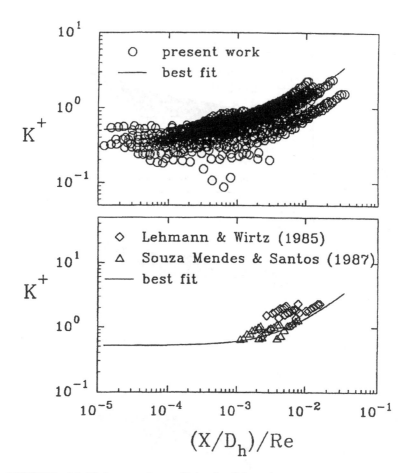

FIGURE 9.   Modified pressure drop coefficient for all data points.

gion of an array of rectangular blocks when $400 \leq Re \leq 15000$, B/L = 0.5, 0.125 $\leq$ S/L $\leq$ 0.5, and 0.125 $\leq$ H/L $\leq$ 1.5.

To compare this correlation with the literature, the pressure results of Lehmann and Wirtz [1985] and those of Souza Mendes and Santos [1987] are shown in the lower graph. There are a number of differences between these investigations. Lehmann and Wirtz studied the problem for B/L = 0.25, $0 \leq$ S/L $\leq$ 1, and 0.25 $\leq$ H/L $\leq$ 0.75, which are somewhat different from the geometric parameters of the present study. In addition, their test section geometry was two dimensional. The pressure results reported by Souza Mendes and Santos [1987] are for B/L = 3/8, S/L = 0.25, and H/L = 5/8. Despite all the differences, the level of agreement seen in Figure 9 is relatively good.

A CORRELATION FOR HEAT
TRANSFER AND WAKE EFFECT

THE EXPERIMENTAL SETUP AND PROCEDURE

The experimental apparatus is very similar to that of pressure drop. It consists of a bell-mouth inlet, a flow development section (559 mm), test section (343 mm), a flow redevelopment section (591 mm), a venturimeter, a control valve, and a blower. The test section and the corresponding upstream and downstream ducts have a rectangular cross section. The width of the cross section is 178 mm, but the height, i.e., H + B in Figure 10a, depending on the dimensions of the array, varies from 19.1 to 50.8 mm. The laboratory air is drawn into the apparatus through the bell-mouth inlet, and after flowing through the test section (Figure 1) and the related flow passages, it exits the flow circuit through the blower operating in suction mode.

The air flow is adjusted by the main control valve. There is also available a bypass valve that is used for fine adjustments. The flow rate is measured by a precalibrated venturimeter that is located downstream from the test section and before the main control valve.

The rectangular blocks of the test section are made of copper and are positioned along the lower wall in an in-line arrangement. The geometry of the test section, as noted in Figure 10, is identified with S, H, L, and B. In this investigation, S ranged from 6.4 to 8.4 mm, H from 6.4 to 25.4 mm, L from 25.4 to 49.8 mm, and B from 12.7 to 25.4 mm. From the combination of these dimensions, seven test sections were fabricated that could readily be identified with their respective dimensionless geometric parameters B/L = 0.5; S/L = 0.128 and 0.33; and H/L = 0.128, 0.25, 0.255, 0.50, 0.75, 0.765, and 1.0. With this arrangement, the number of copper rectangular blocks along and across a given test section was either 5×3 or 8×5. In addition, a number of dummy plexiglass blocks of the same dimensions were positioned downstream of the last copper block to eliminate possible extraneous effects. The air pressure is measured by pressure taps along the wall of the test section. Details of the location of pressure taps and the pressure drop correlations were reported in the preceding section.

As shown in Figure 10c, a thermofoil heater is attached to the bottom of the copper block. To minimize the heat losses, a layer of silica powder was placed at the bottom between the thermofoil heater and the plexiglass lower wall. This arrangement was repeated for all copper blocks of the array. Moreover, the outside surfaces of the test section, including the upstream and downstream ducts, were completely wrapped and insulated by 100 mm of common glass wool insulation.

Before the onset of a data run, the proper test section is selected and assembled. Then, to prevent air from leaking into the test section, all suspected joints are sealed by silicon rubber and thoroughly tested for leaks with the aid of soap solution. Subsequently, the flow and power to thermofoil heater are

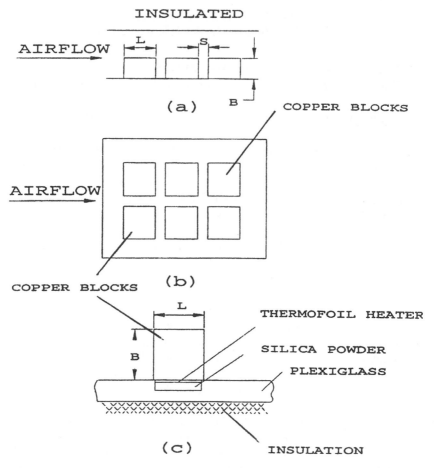

FIGURE 10. Schematic of the test section, (a) side view, (b) top view, (c) a typical heated block.

turned on and adjusted for a predetermined Reynolds number. It typically takes 30 minutes to warm up the apparatus.

In a typical data run, only one copper block is heated at a time and the remaining blocks are adiabatic. The input power of the thermofoil heater is adjusted at such a level that the temperature difference between the block and ambient remains constant. This temperature difference varies from 30 to 50°C depending on the run. Once the heated block reaches the steady state, the various parameters are recorded. These parameters are block surface temperature (the average temperature of three surface-mounted thermistors), inlet air temperature (typically 20 to 25 ±0.1°C), air temperature downstream of the test section (usually 1°C higher than the inlet temperature, ±0.1°C), pressure drop across the venturimeter (0.05017 to 9.80 ±1 × 10⁻⁵ mm Hg), inlet pressure of the venturimeter relative to ambient (0.01 to 13 ±1 × 10⁻⁵ mm Hg), ambient barometric pressure (740 to 770 ±0.1 mm Hg), thermofoil heater voltage (5 to

35 ±0.3 V), and resistance of the thermofoil heater (30 to 32 ±0.3 Ω). Except for the barometric pressure, all other pressures were measured by a pressure transducer and an electronic manometer. All pressure and temperature readings were performed by an IBM PC. The computer scanned the temperature and pressure signals at a scan rate of (50/22) per second.

DATA REDUCTION

The convective heat transfer coefficient is evaluated from, $h = \dot{Q}/[A(T_h - T_{in})]$, where $\dot{Q} = V^2/R$ – (conduction losses + radiation losses). In these equations, $\dot{Q}$ is the rate of heat convection from the heated block to air stream, A is surface area of the block, T_h mean surface temperature of the block, T_{in} inlet air temperature, V voltage across the thermofoil heater, and R is the thermofoil electrical resistance.

Conduction losses were evaluated from a two-D and three-D numerical simulation of heat transfer in the copper block and the adjacent walls. However, due to low thermal conductivity of plexiglass, the effect of longitudinal conduction through the base board was negligible. Radiation losses were estimated from a simplified model where the heated block is treated as a small radiating object surrounded by a large environment. The results suggested that the conduction losses were 10% of V^2/R and the radiation losses, with an emissivity value of 0.15 for copper, were always less than 1% of V^2/R. The h values were subsequently nondimensionalized and expressed in terms of Nusselt number with the conventional definition, $Nu = hL/k$.

The heat transfer results will be presented in terms of Reynolds number, $Re = VH/v$, based on mean air velocity, V, in the bypass channel above the blocks and the dimension H (Figure 10a). The velocity V is obtained by dividing the air volume flow rate, as measured by the venturimeter, by W × H (W = 178 mm).

UNCERTAINTY ANALYSIS FOR HEAT TRANSFER DATA

The parameters recorded during the calibration process and the respective uncertainties are: barometric pressure, ±0.1 mm Hg; venturimeter inlet gage pressure, ±10^{-5} mm Hg; pressure drop across the venturimeter, ±10^{-5} mm Hg; pressure drop across the laminar-flow element, ±10^{-5} mm Hg; and inlet air temperature, ±0.1°C. In addition to these values, the uncertainty of laminar-flow element is estimated as ±1% of volume flow rate.

Other variables used to evaluate the final uncertainties are: duct width, ±0.1 mm; dimension H (Figure 10a), ±0.1 mm; block height B, ±0.1 mm; inlet and throat diameters of venturimeter, ±0.1 mm; test section gage pressure, ±10^{-5} mm Hg; block surface temperature, ±0.1°C; thermofoil voltage, ±0.3 V, and thermofoil resistance, ±0.3 Ω. Using the guidelines of Kline [1985] and Abernethy et al. [1985], the mean uncertainty of Nu is estimated as 4.8%, with the maximum uncertainty being 7.9%. The uncertainties of other parameters are given in the figure captions.

HEAT TRANSFER RESULTS AND DISCUSSION

The entrance region adiabatic heat transfer coefficients are presented in Figures 11 and 12. The ordinate is nondimensionalized with the periodic fully developed Nusselt number, Nu_{fd}, at the Reynolds number corresponding to the same set of data. The coordinate X is measured along the flow with X = 0 corresponding to the leading edge of the first block. The family of curves in Figure 11 are for S/L = 0.128 and H/L ranges from 0.128 to 0.765, while those in Figure 12 are for S/L = 0.33 and H/L = 0.25 to 1.0. It is evident from the figures that the distribution of Nu/Nu_{fd} is almost independent of Reynolds number and H/L. A careful examination of the numerical value of the individual data points show that the mean deviation of data from the best fit (solid lines) is 5.3%, and the maximum deviation does not exceed 10.4%.

The family of curves seen in Figures 11 and 12 show a common trend. Starting with a relatively large value, the Nusselt number decreases rapidly and approaches the fully developed value. This trend is reminiscent of heat transfer coefficient in the entrance region of a duct. However, in addition to the high temperature gradients at the entrance, there exists a strong flow separation zone on the first block of the array (Figure 2 and Molki et al. [1993]). It is well known that the points of flow reattachment have higher transfer coefficients (e.g., see Molki and Hashemi [1992]). Therefore, this flow separation and the subsequent flow reattachment has an enhancing effect on the block-averaged heat transfer coefficients reported in Figures 11 and 12. It should be noted that the extent of enhancement is even more pronounced at special locations such as the line of reattachment.

The data reported in Figures 11 and 12, a total of 176 points, are presented in Figure 13. Also shown in this figure are the results of Anderson and Moffat [1990] for H/L = 0.102 – 0.735, S/L = 0.273, and B/L = 0.204. Despite all the differences between the two investigations, it is seen that all data follow a certain trend. This suggests that a curve fit to all data may be a proper way of representing the Nusselt number in the entrance region of the array. The solid line in Figure 13 is the least squares best fit to the data with the equation

$$\frac{Nu}{Nu_{fd}} = 1 + 0.0786 (\frac{X}{D_h})^{-1.099} \tag{8}$$

The mean and maximum deviation of data from this equation are, respectively, 4.8% and 19.8%, while the mean and maximum experimental uncertainty is estimated as 16.1% and 44.2%. These figures suggest that Equation 8 is a possible correlation for Nu/Nu_{fd}. According to this equation, if the $Nu/Nu_{fd} = 1.01$ marks the beginning of the periodic fully developed region, then the entrance region of the array is identified with $0 \le X/D_h \le 6.53$. Therefore, Equation 8 is particularly useful when $X \le 6.53\ D_h$. Otherwise, the fully developed results can be used.

To facilitate the use of Equation 8 in practical applications, the periodic fully developed Nusselt numbers had to be written in a convenient way. After extensive examination of data and several curve-fitting attempts, a modified

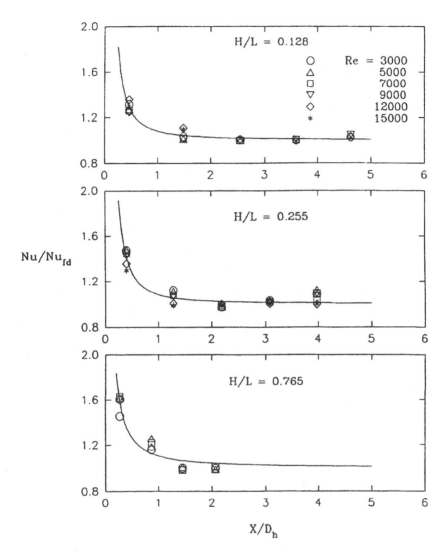

FIGURE 11.   Entrance adiabatic heat transfer coefficients for S/L = 0.128 (mean and maximum uncertainty = 16.1%, 44.2%).

Nusselt number emerged as $Nu_{fd}^+= 2.819\, Nu_{fd}\, Re^{-0.607}\, (S/L)^{-0.295}$, which is presented in Figure 14. Also seen in this figure are the results from the literature. The figure suggests that the modified Nusselt number is a suitable parameter that correlates the data fairly well. The correlation is obtained from a least squares curve fit to all data with the equation

$$Nu_{fd}^+ = 0.968\left(\frac{H}{L}\right)^{-0.670} \tag{9}$$

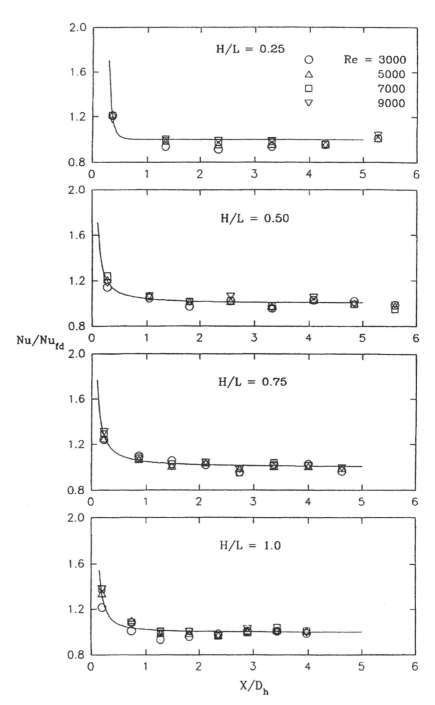

FIGURE 12. Entrance adiabatic heat transfer coefficients for S/L = 0.33 (mean and maximum uncertainty = 16.1%, 44.2%).

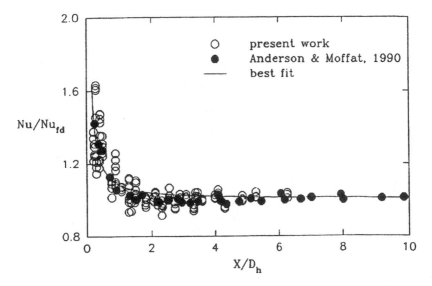

FIGURE 13. Entrance adiabatic heat transfer coefficients for all data and comparison with literature (mean and maximum uncertainty = 16.1%, 44.2%).

The mean and maximum deviation of data from Equation 9 are, respectively, 12.8% and 41.1%.

There are a number of differences between the present results and those of the literature as seen in Figures 13 and 14. Anderson and Moffat [1990] performed their experiments for H/L = 0.102 to 0.735, S/L = 0.273, and B/L = 0.204, while those of Sridhar [1990] were H/L = 0.5 to 5.5, S/L = 0.33 to 0.5, and B/L = 0.5. Sparrow et al. [1982] did not use the electrically heated blocks in their experiments, instead, they employed a mass transfer technique and the heat-mass analogy to obtain the heat transfer coefficients. Their experiments were conducted for H/L = 0.625, S/L = 0.25, and B/L = 0.375. On the other hand, Hollworth and Fuller [1987] results in Figure 14 are measured for a staggered array of blocks with H/L = 0.25 and 0.5, S/L = 1.0, and B/L = 0.25. Despite all these differences, it is noteworthy that the modified Nusselt number has correlated the data well. Therefore, Equations 8 and 9 are suggested as a complete set of correlations for the adiabatic heat transfer coefficient in array of heated blocks.

Adiabatic temperature of the first block situated immediately behind the heated block is presented in Figure 15 as a function of Re with H/L and S/L as parameters. This dimensionless temperature, or the so-called wake effect, is defined as

$$\theta_1 = \frac{T_{al} - T_{in}}{T_h - T_{in}} \tag{10}$$

In this equation, T_{al}, T_{in}, and T_h are, respectively, the adiabatic (unheated) temperature of the first block, the inlet air temperature, and the temperature of the

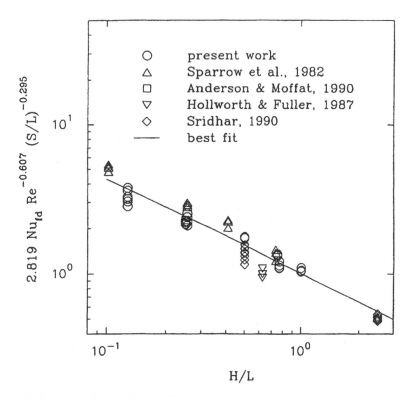

FIGURE 14. Distribution of the modified periodic fully developed Nusselt number (mean and maximum uncertainty = 7.8%, 15.0%).

heated block. It is evident from the figure that the effect of H/L on θ_1 is negligible. However, the effect of the interblock spacing, S/L, is somewhat important. In fact, increasing S/L shifts the location of the first adiabatic block further away from the thermal wake of the heated block and, therefore, the temperature is reduced. The results of Arvizu and Moffat [1982] for S/L = 2.0 and 3.0, Wirtz and Dykshoorn [1984] for S/L = 1, and Sridhar et al. [1990] for S/L = 0.33, are also consistent with this observation.

By defining the modified wake effect according to

$$\theta_1^+ = 2.680 \ \theta_1 \ Re^{0.168} \tag{11}$$

the results are brought together in Figure 16. In this figure, the solid line is the curve fit to all data with the equation

$$\theta_1^+ = 0.998 \left(\frac{S}{L}\right)^{-0.540} \tag{12}$$

The mean and maximum deviation of data from the solid line are, respectively, 6.3% and 34.4%. The uncertainty analysis indicates that the mean and maxi-

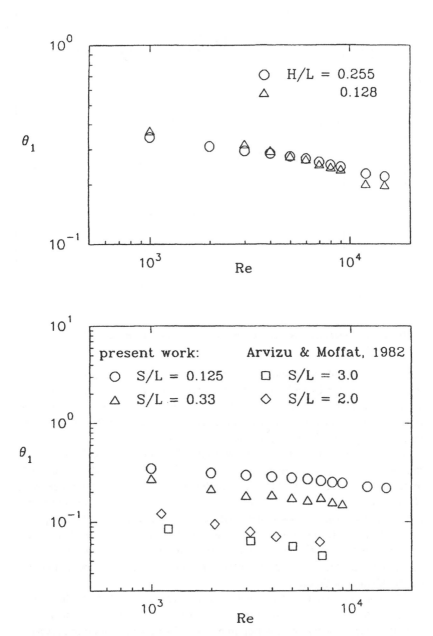

FIGURE 15. Thermal wake effect for the first adiabatic block (mean and maximum uncertainty = 1.4%, 2.8%).

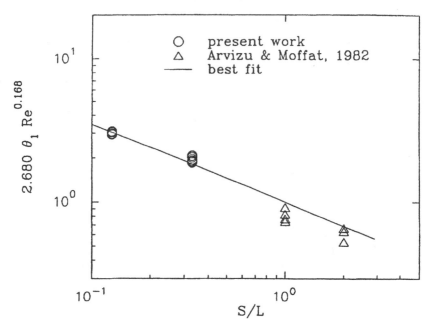

FIGURE 16. Modified wake function (mean and maximum uncertainty = 28.3%, 30.1%).

mum experimental uncertainty of θ† are 28.3% and 30.1%, respectively. Since the mean deviation is well within the mean uncertainty and the maximum deviation and uncertainty have comparable values, it may be suggested that Equation 11 be considered for the range $0.125 \leq S/L \leq 3.0$, which extends beyond the range of S/L in the present investigation.

The wake effect for other downstream blocks is presented in Figure 17. In this figure, θ_N is defined as $\theta_N = (T_{aN} - T_{in})/(T_h - T_{in})$. Here, the subscript N refers to the N*th* adiabatic block situated downstream of the heated block, with N = 0 corresponding to the heated block itself. During the wake effect experiments, the heated block was located at the centerline and at either the first (leading) or at the third row of the array. The experimental runs were also carried out for different geometries. As evidenced from the data in Figure 17, the location of the heated block and the geometry does not have a marked effect on the distribution of θ_N/θ_1.

There are a number of investigators who have correlated the θ_N/θ_1 data as 1/N (e.g., Arvizu and Moffat [1982], Moffat et al. [1985], and Sridhar et al. [1990]) or $(1/N)^m$ where m is a function of Re (Wirtz and Dykshoorn [1984]). The major difficulty with this correlation is that when the number of blocks is large ($N \to \infty$), 1/N approaches zero. In an insulated duct with a number of heated blocks, the temperature of the downstream adiabatic blocks approaches the mean temperature of the air flow, which is somewhat higher than the inlet temperature. Therefore, a suitable correlation is the one which incorporates this

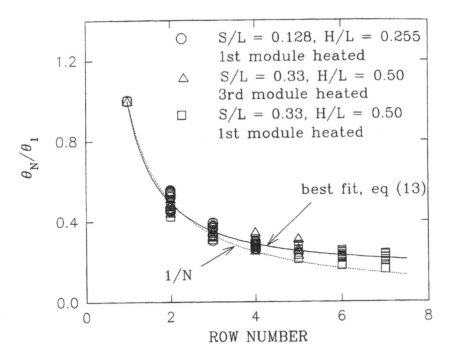

FIGURE 17. Thermal wake effect for other downstream blocks (mean and maximum uncertainty = 4.0%, 44.0%).

limiting case. In the light of the foregoing discussion, the wake effects are correlated as

$$\frac{\theta_N}{\theta_1} = 0.151 + 0.849 \, N^{-1.314} \qquad (13)$$

Figure 17 shows that Equation 13 is a better fit to data than $1/N$.

Attention is next turned to the wake effect on the blocks situated in the flanking columns of the heated block. In Figure 18, the ordinate, θ_{fN}/θ_N is the ratio of the wake effect for the flanking block at the same row, i.e., $\theta_{fN} = (T_{fN} - T_{in})/(T_h - T_{in})$, to θ_N defined earlier. It is seen from the figure that farther away from the heated block θ_{fN} increases with respect to θ_N and is independent of Re. In the range of $N = 1$ to 4, θ_{fN} varies from 5.8% to 27.5% of θ_N.

The wake effect for the flanking column may be represented by

$$\frac{\theta_{fN}}{\theta_N} = 0.0575 \, N^{1.128} \qquad (14)$$

which is a least-squares fit to all data. The mean and maximum deviation of data from this equation are, respectively, 7.8% and 22.6%, while the mean and maximum uncertainty are 13.0% and 16.0%.

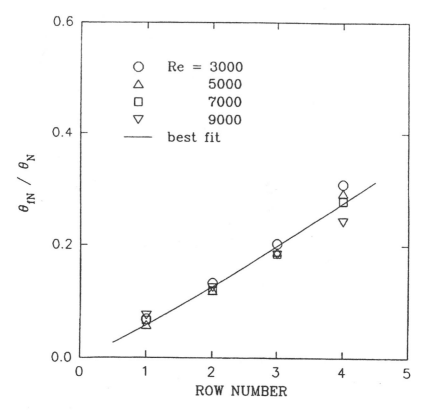

FIGURE 18. Thermal wake effect for the blocks in the flanking columns (mean and maximum = 13.0%, 16.0%).

Next, attention is turned to the application of the above correlations for predicting the temperature of the circuit-board elements. The application is limited to the geometry and the range of parameters considered in this paper. However, they can also be used as first approximation for a wider range of geometry and parameters in practical problems.

APPLICATION OF THE PROPOSED CORRELATIONS

Based on the aforementioned correlations, a marching procedure is proposed for estimating the operating temperature of elements in a circuit-board with random heating using the superposition approach (Arvizu and Moffat [1982]). In this procedure, the temperature of each block is considered to be composed of two components. One is due to self-heating ($\Delta T_{sh} = T_h - T_a$), the other is due to the thermal wake of the upstream heated blocks (ΔT_{we}). It should be noted that every downstream block is affected by the thermal wake of every upstream block situated in the same column or in the flanking columns. Therefore, the thermal wake component will be evaluated from the summation of the wake effects of all the upstream heated blocks.

The algorithm is started by providing the geometrical parameters (number of rows and columns, L, B, S, H, W), air inlet velocity, V_{in}, inlet air temperature, T_{in}, and power dissipation in each block. These values are used to obtain S/L, H/L, X/D_h, and Re. Then, the temperatures are computed according to the following steps:

1. Adiabatic heat transfer coefficient, h, is obtained from Equations 8 and 9 for all blocks of the array.
2. Operating temperatures of the blocks in the first row are evaluated from, (Power – heat losses) = $hA(T_h - T_a)$, where Power is the heat generation within the block. It should be noted that the adiabatic temperature of the blocks situated in the first row is equal to the inlet air temperature, i.e., $T_a = T_{in}$.
3. Wake effect of the heated blocks on all downstream blocks situated in the same column, i.e., $\Delta T_{we,s} = T_{aN} - T_{in}$, is obtained from Equations 10–13. In this way, for every downstream block a different $\Delta T_{we,s}$ value is evaluated and assigned to that block.
4. Wake effect of the heated blocks on the flanking columns is evaluated from Equation 14. Only one column to the left and one to the right of the respective heated block is considered. The wake effect on other columns has been found to be negligible. In this manner, two additional wake effects, namely $\Delta T_{we,l}$ and $\Delta T_{we,r}$, are found and assigned to the block. In this procedure, for heated blocks in the first row, three $\Delta T's$ are assigned to each of the blocks situated downstream.
5. Adiabatic temperature of the blocks in the second row are estimated from $T_a = \Delta T_{we} + T_{in}$, where $\Delta T_{we} = \sum \Delta T_{wei}$, and i = s,l,r.
6. Consider the second row of heated blocks. Using the adiabatic temperatures of the previous step, repeat steps 2 to 5 to find the operating temperatures of the blocks in the second row and their respective wake effects on downstream blocks. At the end of every cycle of iteration, the adiabatic temperature of the blocks in the next row is obtained from all the previously obtained $\Delta T_{we,i}$'s according to step 5.

A USER-FRIENDLY PROGRAM FOR PREDICTION OF ARRAY TEMPERATURE

The aforementioned procedure is incorporated into a user friendly FORTRAN program that is available from one of the authors (MF) upon request. A sample output from this computer program is shown in Figure 19. The figure shows the top view of the array. The cross-hatched blocks are heated while the remaining blocks are adiabatic. The first of the four numbers on the heated blocks is the power generation in watts. This number is not seen for the un-

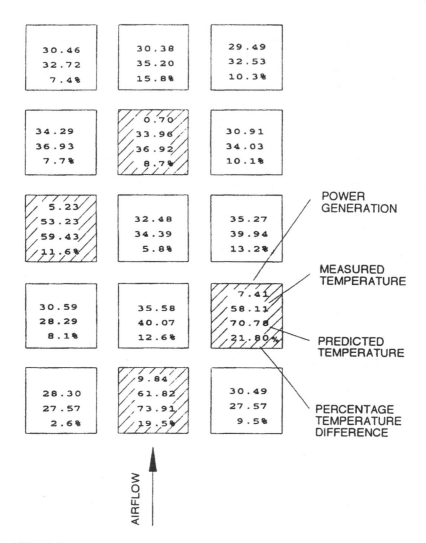

FIGURE 19. A typical temperature prediction for the elements of a circuit board. The data are for L = 49.8 mm, B = 25.4 mm, S = 6.4 mm, H = 38.1 mm, V_{in} = 1.24 m/s, and T_{in} = 27.57°C, corresponding to S/L = 0.128, H/L = 0.765, B/L = 0.5, and Re = 5000

heated blocks. The last three numbers on the heated and unheated blocks indicate the measured and computed temperatures in degree C, and their difference, respectively. Examination of difference percentages indicates that the difference between experiment and prediction is in the range from 2.6 to 21.8%, and the mean difference is 11.0%.

CONCLUSION

The research reported here is an experimental investigation of heat transfer and pressure drop in the entrance region of an in-line array of heated blocks. The study is aimed at obtaining correlations for pressure drop, adiabatic heat transfer coefficient, and thermal wake effects so that the operating temperature of the real circuit boards with a similar geometrical layout can be estimated. However, since the real electronic components are irregular both in placement and size, the results reported in this paper may not be applicable to highly irregular layouts. In all of the experiments the working fluid was air.

Flow visualization indicated the presence of a highly separated flow on top of the first module followed by flow reattachment and recirculation. It was further found that flow reattachment is curved, and thus the flow field is thoroughly three dimensional. Farther downstream, the flow visualization could not present a clear evidence of flow separation. However, the pressure results indicate that at higher values of Re, the flow is possibly separated behind the downstream modules.

Pressure drop results are presented and correlated through the definition of a modified pressure drop coefficient. This correlation is good to within ±54.4% and is a convenient tool for estimating the pressure drop in arrays of rectangular blocks.

The entrance heat transfer coefficients revealed that Nu/Nu_{fd} is a function of X/D_h, and its dependence on geometrical parameters and Reynolds number is within the limits of the experimental uncertainty. These coefficients were successfully correlated by defining a modified Nusselt number. The correlation is obtained by a curve-fitting procedure that includes the data from other investigators, and thus is expected to be valid over a wider range than that of the present work.

Temperature measurements indicated that the thermal wake effect, θ_1, for the first adiabatic block behind the heated block is a function of Re and S/L and is insensitive to H/L. A modified wake effect, θ_1^*, absorbed the effect of Re and correlated the results as a function of S/L.

The wake effect for other blocks, when presented as θ_N/θ_1, are found to be independent of the position of the heated block and geometry. It was further found that the commonly employed 1/N function is not a well representation of θ_N/θ_1 distribution. Instead, it was correlated as $0.151 + 0.849 \, N^{-1.314}$, which gives a more realistic asymptotic value for θ_N/θ_1 when $N \rightarrow \infty$. Correlations were also found for the blocks situated in the flanking columns.

The present correlations are incorporated into a FORTRAN program to predict the operating temperatures of the circuit boards of similar geometry with random heating. The results indicate that the temperatures can be predicted with sufficient accuracy for practical application.

APPENDIX A: NOMENCLATURE

A_1 inlet cross-sectional area of venturimeter, $(\pi/4)D_1^2$
B   module height
C_D flow coefficient of venturimeter, $C_D = Q[(D_1/D_2)^2 - 1]^{0.5}/[A_1\,(2\Delta P/\rho_1)^{0.5}]$
D_h hydraulic diameter, mm, $2W(H + B)/(W + H + B)$
D_1 inlet diameter of venturimeter
D_2 throat diameter of venturimeter
f_{2H} friction factor
h   mean heat transfer coefficient for a rectangular block
H   distance between top surface of modules and the opposite wall
K   pressure drop coefficient
K^+ modified pressure drop coefficient
L   plane dimension of square module
Nu   Nusselt number, hL/k
P_i air pressure at the ith pressure tap
P_o air pressure at the first pressure tap
Q   volume flow rate
$\dot{Q}$ rate of heat convection
R   thermofoil electrical resistance
Re   Reynolds number, VH/v
S   intermodule spacing
T   temperature, °C
V   voltage across the thermofoil heater, air mean velocity in the bypass channel
W   duct width, W = 178 mm
X   axial coordinate, X = 0 at the leading edge of the first block

GREEK SYMBOLS
θ wake effect
v kinematic viscosity
ρ density

SUPERSCRIPTS
+   modified parameter

SUBSCRIPTS
al   refers to adiabatic temperature of the first block behind the heated block, Equation 5
aN   refers to adiabatic temperature of the Nth block behind the heated block
fd   refers to periodic fully developed, Equation 4
fN   refers to flanking column, Equation 9
h   refers to heated block, Equation 5
in   refers to inlet condition, Equation 1

N refers to the N*th* block behind the heated block, Equation 8
sh refers to self heating
we refers to wake effect

REFERENCES

Abernethy, R.B., Benedict, R.P., and Dowdell, R.B. 1985. ASME measurement uncertainty, *ASME J. of Fluids Eng.*, 107:161–164.

Anderson, A.M., and Moffat, R.J. 1990. A new type of heat transfer correlation for air cooling of regular arrays of electronic components, in *Proc. of ASME Winter Annu. Meet.*, 27–39.

Anderson, A.M., and Moffat, R.J. 1991. Direct air cooling of electronic components: Reducing component temperatures by controlled thermal mixing, *ASME J. of Heat Transfer*, 113:56–62.

Arvizu, D.E., and Moffat, R.J. 1982. The use of superposition in calculating cooling requirements for circuit board mounted electronic components, in *Proc. of the 32nd Electron. Components Conf.*, IEEE, 32:133–144.

Asako, Y., and Faghri, M. 1988. Three-dimensional heat transfer and fluid flow analysis of arrays of square Blocks encountered in electronic equipment, *Numerical Heat Transfer*, 13:481–498.

Asako, Y. and Faghri, M. 1991. Parametric study of turbulent three-dimensional heat transfer of arrays of heated blocks encountered in electronic equipment, *Heat Transfer in Electron. Equipment*, HTD, 171:135–141.

Chou, J.H. and Lee, J. 1988. Reducing flow nonuniformities in LSI packages by vortex generators, in *Cooling Technology for Electronic Equipment*, W. Aung, Ed., Hemisphere, Washington, D.C., 113–124.

Faghri, M., Molki, M., Chrupcala, J., and Asako, Y. 1995. Entrance analysis of turbulent flow in an array of heated rectangular blocks, IEEE Transactions on Components, Packaging, and Manufacturing, Part A, Vol. 10, No. 3.

Faghri, M., Ray, A., and Sridhar, S. 1991. Entrance heat transfer correlation for air cooling of arrays of rectangular blocks, *Heat Transfer Enhancement in Electronics Cooling*, ASME HTD 183:19–23.

Garimella, S.V. and Eibeck, P.A. 1992. Onset of transition in the flow over a three-dimensional array of rectangular obstacles, *J. of Electron. Packag.*, 114:251–255.

Hollworth, B.R. and Fuller, H.A. 1987. Heat transfer and pressure drop in a staggered array of air-cooled components, in Proc. of the Int. Symp on Cooling Technol. for Electron. Equipment, Honolulu, 732–748.

Kang, S.S. 1992. The thermal wake function for rectangular electronic modules, *National Heat Transfer Conf.*, Open Forum, San Diego, CA.

Kline, S.J. 1985. The purposes of uncertainty analysis, *ASME J. Fluids Eng.*, 107:153–160.

Lehmann, G.L. and Wirtz, R.A. 1985. The effect of variations in stream-wise spacing and length on convection from surface mounted rectangular components, ASME HTD, 48:39–47.

Moffat, R.J. and Anderson, A.M. 1988. Applying heat transfer coefficient data to electronics cooling, ASME Winter Annu. Meet, Chicago, IL.

Moffat, R.J., Arvizu, D.E., and Ortega, A. 1985. Cooling electronic components: forced convection experiments with an air-cooled array, *Heat Transfer in Electronic Equipment*, ASME HTD 48:17–27.

Molki, M., Faghri, M., and Ozbay, O. 1993. A new correlation for pressure drop in arrays of rectangular blocks in air-cooled electronic units, *Natural and Forced Convection*, ASME J. of Fluids Eng., 116:856–886.

Molki, M. and Hashemi, A. 1992. Turbulent convective mass transfer downstream of a perforated baffle blockage, *Int. J. Heat and Fluid Flow*, 13(2):116–123.

Peterson, G.P. and Ortega, A. 1990. Thermal control of electronic equipment and devices, in *Advances in Heat Transfer*, Vol. 20, J.P. Hartnett and T.F. Irvine, Jr., Eds, Academic Press, New York.

Souza Mendes, P.R. and Santos, W.F.N. 1987. Heat-transfer and pressure drop experiments in air-cooled electronic-component arrays, *J. Thermophys.*, 1:373–378.

Sparrow, E.M., Molki, M., and Chastain, S.R. 1981. Turbulent heat transfer coefficients and fluid flow patterns on the faces of a centrally positioned blockage in a duct, *Int. J. Heat Mass Transfer*, 24:507–519.

Sparrow, E.M., Niethammer, J.E., and Chaboki, A. 1982. Heat transfer and pressure drop characteristics of arrays of rectangular modules encountered in electronic equipment, *Int. J. Heat Mass Transfer*, 25:961–973.

Sparrow, E.M., Vemuri, S.B., and Kadle, D.S. 1983. Enhanced and local heat transfer, pressure drop, and flow visualization for arrays of block-like electronic components, *Int. J. Heat Mass Transfer*, 26:689–699.

Sridhar, S. 1990. Heat transfer and fluid flow behavior in arrays of rectangular blocks encountered in electronic equipment, Master's thesis, University of Rhode Island.

Sridhar, S., Faghri, M., and Lessmann, R.C. 1990. Heat transfer behavior including thermal wake effects in forced air cooling of arrays of rectangular blocks, ASME HTD, 153:15–26.

Tai, C.C. and Lucas, V.T. 1985. Thermal characterization of a card-on-board electronic package, *Heat Transfer in Electronic Equipment*, ASME HTD, 48:49–57.

Wirtz, R.A. and Dykshoorn, P. 1984. Heat transfer from arrays of flat packs in a channel flow, in *Proc. of 4th Annu. Int. Electron. Packag. Soc.*, Baltimore, 247–256.

Wirtz, R.A. and Chen, W. 1991. Laminar-transitional convection from repeated ribs in a channel, *Heat Transfer in Electronic Equipment*, ASME HTD, 171:89–94.

Chapter 3

FORCED AIR COOLING
OF LOW-PROFILE PACKAGE ARRAYS

R. A. Wirtz

CONTENTS

0-8493-9447-3/96/$0.00+$.50
© 1996 by CRC Press, Inc.

81

INTRODUCTION

Air continues to be the preferred coolant for most electronics installations. Direct air cooling, where fan-driven, conditioned air is passed directly over board-mounted packages, is a convenient and popular cooling scheme. If circuit boards are rack mounted, the space between boards acts as a conduit to direct the coolant over and around modules, allowing for convective cooling of both the modules and the pc-board.

In these applications, heat flows from the device to the ultimate sink via a complicated path. On the one hand, heat is conducted directly to the package surface, where it is removed by convection to the coolant and radiation to the surroundings. The remainder of the energy is conducted through the mounting substrate and attachment leads to the pc-board where it is either conducted to the system enclosure, or convection processes on the back of the board or between modules remove it. A complete thermal analysis to predict device operating temperatures would require consideration of all of the heat transfer paths described above. However, if the convection mechanisms are suitably quantified, the analysis can be completed using a standard conduction analyzer in conjunction with separate calculations for the convection and radiation components of the heat transfer process.

The objective in this chapter is to describe the convection process occurring in a class of air-cooled packaging configurations where conduction and radiation heat transfer are absent. Heat transfer correlations and calculation procedures that can be used to estimate package surface temperatures and coolant flow pressure drops are presented.

ARRAY GEOMETRY

Packaging configurations are diverse, usually including a mix of package sizes on a given pc-board. In addition, some packages may include heat sinks, and electrical cabling or card separators may partially obscure the flow. In order to develop a systematic description of convection in electronics, we restrict our attention to unobstructed flow over uniform arrays of equal-sized elements. Where information regarding the effect of array nonuniformity is available, it is pointed out.

Consider a uniform, in-line array of modules mounted on a pc-board, as shown in side view in Figure 1. Let x and z denote the stream-wise and cross-stream directions, and y is the coordinate perpendicular to the plane of the figure. Each array element has height, a, length, L_x, and width L_y. The corresponding spacing between elements is c_x and c_y. We can define the array (area) packaging density as

$$D = \frac{L_x L_y}{S_x S_y} \tag{1}$$

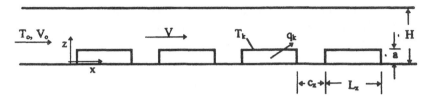

FIGURE 1.   Side view of a uniform, in-line array of convectively cooled electronic packages.

where $S_x = L_x + c_x$ and $S_y = L_y + c_y$ are the packaging pitch in x and y directions, respectively. The packaging density can range from zero to unity. For square packaging arrangements ($L_x = L_y$, $c_x = c_y$) the packaging density is given by $(L/S)^2$. Package lengths range from about 10 mm to 80 mm, and package aspect ratios (L/a) can range from 2 to 20. Package spacing, c, is typically 4 mm – 16 mm. Array densities typically range from 0.25 to 0.9.

Package dimensions and spacings, in conjunction with the pc-board spacing and array size, comprise the principal length scales that are known to influence cooling performance. The effect of other geometric features, such as package edge radii, the presence of lead attachments, etc., are assumed to have only a small impact on convection.

CONVECTION PROCESSES

As shown in Figure 1, the coolant (air) at temperature, T_c, and average velocity, V_c, flows into a channel of height, H. Such a channel might be formed by two adjacent pc-boards, or by a board bounded by the system enclosure. Array elements have surface temperature, T_k, and the rate of convective heat removal is q_k, where k is the row number in the array. Upon encountering the array, the entry flow divides. Part is diverted to the space above the modules (the *bypass flow*), and the remainder is channeled into the spaces between array elements (the *array flow*). The extent of interaction between the bypass flow and the array flow is not generally known. It has been shown for regular, in-line arrays containing cubical array elements (where the element height, a, is not negligible) that the fluid velocity in the array is significantly reduced relative to that in the by-pass region [Arvizu and Moffat, 1982]. Under these circumstances, convective cooling of the board surfaces between packages probably plays only a secondary role in package cooling. On the other hand, board-convection is probably a significant part of the overall heat transfer problem when package heights are very small, such as in surface-mount applications.

The methods of analysis that follow are directed at geometric situations where the package height is sufficiently large that the coolant flow is modified away from the "smooth duct" condition. Situations where a is *small* can be analyzed using conventional superposition techniques [Kays and Crawford, 1993] in conjunction with heat transfer correlations suggested by Lehmann and Pembroke [1991a].

A good description of the bypass flow is given by the local average velocity, V, which is the average velocity based on the minimum cross sectional area of the flow passage.

$$\frac{V}{V_o} = \frac{1}{1 - \dfrac{aL_y}{HS_y}} \qquad (2)$$

Since both a/H and L_y/S_y are positive and less than unity, Equation 2 shows that the bypass flow is accelerated relative to the entry flow. This is, of course, due to the extra blockage that the array elements present.

Consider air cooling pc-boards spaced 20 mm apart with a fan-driven flow that can range from 200 fpm to 2000 fpm (1 m/s to 10 m/s). The package heights are 4 mm and the packaging density is 60%. Under these conditions the average bypass velocity is roughly 20% greater than the inlet average velocity, and the channel Reynolds number for the bypass flow, Re = $V2(H-a)/v$, will range from approximately 2,400 to 24,000. Thus, we expect convection process, particularly on the tops of the array elements, to have mostly turbulent characteristics.[1] Since the bypass flow and the array flow velocities are of different magnitudes, a shear layer that separates the two regions forms near the tops of the array elements. The shear layer is characterized by relatively high levels of mixing with measured stream-wise turbulence intensities within it approaching 20% or more [Garimella and Eibeck, 1992; Wirtz and Chen, 1992]. This enhanced mixing along the top of the package results in heat transfer rates that are greater than would occur in a smooth-walled channel.

Since the bypass and array flows are different, the convection process on the top of a module is likely to be different from that found on any of the four sides exposed to coolant. As a simplification, we restrict our attention to modules where convection is mostly influenced by the bypass flow since this quantity is relatively easy to quantify. We define a low profile package as one where the package top represents at least 50% of the package heat transfer surface area, i.e.,

$$\frac{L_x L_y}{L_x L_y + 2a(L_x + L_y)} \geq 0.5 \qquad (3)$$

For a square planform package, Equation 3 requires that low-profile packages have aspect ratios, $L/a \geq 4$. This restriction covers the majority of packages found in application.

Wirtz and Mathur [1994] studied the local heat transfer rates on the surfaces of low-profile packages. (They studied square planform packages having $L/a = 6$). These investigators found that the local heat transfer coefficient dis-

---

[1] Quantative measures of laminar-turbulent transition in terms of a transition Reynolds number are not available. An assessment of whether a given flow is laminar, turbulent or transitional is a judgment call.

tribution on the top surface of the package is similar to, but greater than, that expected on a flat plate in a turbulent airstream. They also found that the average heat transfer coefficient on the top surface of these packages is approximately equal to the overall heat transfer coefficient for the package. Therefore, we expect that heat transfer correlations developed for low-profile packages will be appropriately described by the local average velocity, V.

ESTIMATION OF PACKAGE TEMPERATURE

Two convective effects contribute to the temperature rise of an electronic package. The first, called the self-heating effect, is due to heat generated within the package. The second is due to heat releases from other components that are upstream from the package being considered. These upstream heat releases raise the mean temperature of the coolant which then washes over the package in question, raising its temperature. Thus, a completely passive component that resides in the wake of a heat-releasing element will experience a temperature increase. This second heating effect is often called the thermal wake effect.

For forced convection cooling (no buoyancy effects or radiation heat transfer), these two effects may be combined by simple addition (superposition) because the energy equation and its boundary conditions are linear under these conditions. A convenient expression [Arvizu and Moffat, 1982] is

$$T_k - T_o = \left(\frac{q_k}{h_k A_k}\right) + \sum_{i=1}^{k-1} \theta_{k-i} \cdot [T_i - T_o], \quad i < k \tag{4}$$

The first term on the right of Equation 4 expresses the self-heating temperature rise of element k above the entrance temperature, T_o, in terms of its convective heat release rate, q_k, heat transfer surface area, A_k, and the adiabatic heat transfer coefficient, h_k. The term adiabatic is used in describing h_k since it is the heat transfer coefficient observed in the absence of other (upstream) heat releases, and this serves as the quantity's definition.

$$h_k = \frac{q_k}{A_k(T_k - T_o)}\bigg|_{\substack{only \\ q_k \neq 0}} \tag{5}$$

As a consequence, h_k is a function of flow rate, channel and array geometry, fluid properties, and perhaps position in the array. However, its value is not dependent on upstream thermal boundary conditions. This characteristic of h_k allows the two effects (self-heating and upstream heat release) to be separated in Equation 4. From an experimental viewpoint, this is one of the main advantages of the approach since h_k is measured in an array with only the k-th element powered-up and all other elements passive. Moffat and Anderson [1990] discuss the relationship of the adiabatic heat transfer coefficient to other heat transfer coefficient definitions used in heat transfer analysis.

It should also be noted that h_k may be related to the case-to-ambient thermal resistance, R_{c-a}, only where R_{c-a} is reported for conditions of no upstream heating. Under these conditions

$$h_k = \frac{1}{R_{c-a} A_k} \tag{6}$$

The second term on the right of Equation 4 expresses the temperature rise of element k due to thermal wakes from all elements in array rows upstream from k. It contains the thermal wake function, θ_{k-i}, which is the fractional temperature rise of element k due to heat release from element i with $q_k = 0$,

$$\theta_{k-i} = \left.\frac{T_k - T_o}{T_i - T_o}\right|_{\substack{q_i \neq 0 \\ q_k = 0}} \tag{7}$$

Like h_k, θ_{k-i} is a function of flow rate, channel and array geometry, fluid properties, and position in the array. Its experimental determination is relatively straightforward. Experiments are conducted with one array element heated, and the temperature increases of elements downstream from it are recorded.

It should be noted that the form of Equation 4 is appropriate to situations where the upstream heating effect is confined to the same column of the array as the heat release (i.e., thermal wakes do not spread to adjacent columns). Experimental observations on in-line arrays of low profile packages support this assumption [Wirtz and Dykshoorn, 1984]. Furthermore, Moffat and Anderson [1990] have developed the thermal wake concept of Equation 4 in terms of kernal functions, placing the approach on a firm analytical footing.

DETERMINATION OF h_k AND θ

Due to the inherent complexity of the flow, most correlations of h_k, and θ_{k-i} are empirically based. Two approaches have been followed to obtain data used to construct correlations: prototype experiments and model experiments.

PROTOTYPE EXPERIMENTS

Prototype experiments utilize actual package encapsulations mounted to pc-boards using standard attachment methods. A small heating element and temperature sensor replace normal circuitry inside the encapsulation. Package arrays are arranged as they would be in application. The advantage of this approach is that the heat transfer data obtained is a direct measure of the cooling performance of a specific design. The principal disadvantage is that it is difficult, if not impossible, to separate the effect of the different heat transfer modes that are active. This is usually because conduction through attachment leads to the pc-board can not be accurately measured. As a result, heat transfer data obtained in this way is not easily generalized to other, geometrically similar packaging configurations.

MODEL EXPERIMENTS

Model experiments utilize test articles that simulate actual packages. The model packages can be carefully designed to minimize measurement errors, and experimental test rigs can be designed to eliminate unwanted heat transfer modes. For example, guard heaters may be employed to eliminate conduction to the mounting board, and package surfaces can be highly polished to minimize radiation heat transfer. Similitude may then be employed to generalize measurements from a given test rig to other geometrically similar packaging configurations. The disadvantage of this approach is that the model test article may neglect a significant geometric feature contained in the prototype, leading to erroneous conclusions about the convection process.

Copeland [1992] reports prototype measurements that utilized square planform, aluminum/ceramic encapsulated packages (L = 37 mm, L/a = 6.4) that were pin-grid attached to epoxy-glass interconnect boards. Packages were arranged in square arrays with the packaging density, D, ranging from 0.25 to 0.94. The boards, with board-to-board spacing varied over the range $2.2 \leq H/a \leq 5.5$, were tested in a special purpose channel designed to simulate flow through a card rack. For array elements far from the channel entrance, Copeland correlates his heat transfer coefficient data in terms of entrance velocity using a *linear* equation. For a packaging density of D = 0.53 and H/a = 2 he suggests the following equation.[2]

$$h = 20 + 7V_o \tag{8}$$

Wirtz and Mathur [1994] report heat transfer measurements from a *model* experiment that used a packaging configuration similar to the one studied by Copeland. The model experiments utilized machined aluminum blocks that were instrumented with small heaters and calibrated thermocouples. External surfaces were polished to minimize radiation heat transfer. The experiment was designed so that conduction to the mounting board could be accurately measured and subtracted from the heater power to give the convection heat transfer rate. The model packages were larger than those used in the prototype experiment (L = 69.8 mm). However, the package aspect ratio (L/a = 6), packaging density (D = 0.45), and dimensionless channel height (H/a = 2) are almost the same as those used to construct Equation 8.

Figure 2 compares Copeland's linear correlation, Equation 8, with the convection heat transfer data obtained in the model experiment. The prototype measurements of h are much higher than the model experimental results, particularly at low entrance velocities. Part of this difference is attributable to the longer package length employed in the model experiments. The remaining difference is because the prototype measurements mix conduction and convection effects. As the flow rate increases, the convective component increases (the conduction component should be essentially independent of flow rate) and the two results

---

[2] Equation 8 is dimensionally inhomogeneous. This practice is not recommended.

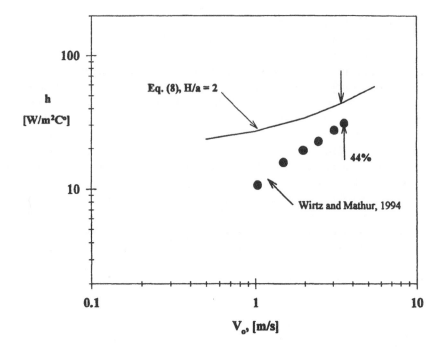

FIGURE 2. Comparison of package heat transfer measurements: Prototype vs. Model experiment.

appear to converge. However, Equation 8 implies the wrong dependence of convection on flow rate. The results are plotted in log–log coordinates in the figure, so the linear equation, Equation 8, plots as a curve that is asymptotic to $h \sim V_o^n$. On the other hand, the model experimental results (which include convection only) appear to plot as a straight line over the entire range of velocities investigated. Thus, the model experiment indicates a logarithmic dependence of the heat transfer rate on coolant velocity, with $h \sim V_o^n$ where n < 1, and this power-law dependence is what has been observed in all reported model experiments.

A regression analysis of the model experiment data leads to the conclusion that $h \sim V_o^{0.8}$. A similar regression analysis of the prototype data, assuming a power-law relation between h and V_o, would give something like $h \sim V_o^{0.5}$. The lower exponent obtained from the prototype measurements is caused by the relatively high heat transfer coefficients reported at low flow rates. These high heat transfer rates are caused by the relatively large contribution that conduction makes to the measured heat transfer at these low flow rates. In general, if conduction and radiation heat transfer is not adequately accounted for in an experiment, the resulting exponent in a power-law correlation of convection heat transfer data will give a low estimate of the power-law exponent.

Since there is some uncertainty regarding the role of conduction and radiation in heat transfer results obtained via prototype experiments, the correlations developed in the following section are limited to results derived from model experiments.

PACKAGE HEAT TRANSFER
UNIFORM, IN-LINE ARRAYS

ADIABATIC HEAT TRANSFER COEFFICIENT

As in any confined flow, fully developed conditions are preceded by a length of channel where the local shear stress and heat transfer coefficient vary in the streamwise direction. Most heat transfer observations of convection from arrays of electronic packages suggest that it takes from three to five package rows for the convection process to become fully developed (in a row periodic sense). For the channel Reynolds numbers generally encountered, the leading rows exhibit higher heat transfer coefficients than the fully developed values. Generally, the leading row h is about 20–30% greater than the fully developed value, and the second-row value is augmented by 5–10%. However, this is only an approximate "rule of thumb." As the Reynolds number decreases the amount of augmentation exhibited by leading-row array elements also diminishes so that at very low Reynolds numbers leading row heat transfer coefficients may be equal to, or lower than, those measured down stream [Garimella and Eibeck, 1990, Lehmann and Pembroke, 1991a]. In the following, we develop a correlation for the fully developed package heat transfer coefficient. Corrections, such as the above mentioned "rule of thumb," can be applied to the fully developed coefficients to account for leading row effects.

Table 1 summarizes the geometrical characteristics of model package arrays considered in some recent heat transfer experiments. The table lists the package length and aspect ratio, packaging density of the array, and dimensionless channel height, H/a. All investigations considered uniform arrays and, except for Anderson and Moffat[3], all considered square planform packages. The package length is seen to vary from 25.4 mm to almost 70 mm and the package aspect ratio varies from 2.67 to 8.75. Packaging density is seen to vary from the

TABLE 1
Geometric Features of Some Model Heat Transfer Experiments

Investigator	L_x [mm]	L/a	D	H/a
Wirtz and Dykshoorn, 1984	25.4	4.00	0.25	1.5–4.6
Sparrow et al., 1982	26.7	2.67	0.64	2.7
Anderson and Moffat, 1992a	37.5	3.95	0.59	1.5–4.6
Wirtz et al., 1994b	56.0	8.75	0.49	1.5–10
Wirtz and Mathur, 1994	69.8	6.00	0.45	2.0
Wirtz and Colban, 1995	69.8	6.00	0.45–0.69	2.0

[3] Anderson and Moffat considered rectangular planform packages with Lx/Ly = 0.81, Sx/Sy = 0.85

relatively sparse array considered by Wirtz and Dykshoorn (D = 0.25) to very dense packaging considered by Wirtz and Colban (D = 0.69). Dimensionless channel heights range from 1.5 to 10.

Figure 3 plots reported heat transfer coefficient measurements of h from the studies listed in Table 1 in terms of the local average velocity, V (see note[4]). The following observations can be made:

1. The data lie in a band having a bandwidth that is about ±25% of any mid-point value.
2. Subsets of the data are generally arranged in parallel straight lines with a decrease in h corresponding to an increase in package length, L.
3. The heat transfer coefficient is seen to increase logarithmetically with velocity, implying a power-law relation, h~V^n, for each subset of the data. Best-fit values of n range from 0.6 to 0.8.

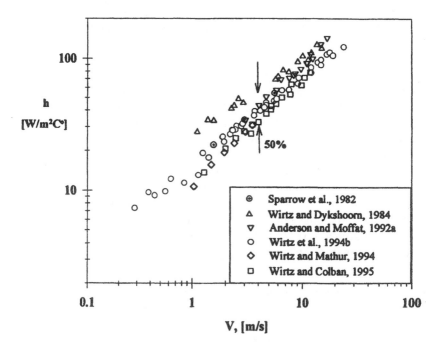

FIGURE 3. Package adiabatic heat transfer coefficient measurements from the model experiments listed in Table 1.

[4] The experimental work of Wirtz et al. [1994]; Wirtz and Mathur [1994]; and Wirtz and Colban [1995] was completed at an altitude of 1300 m M.S.L. In Figure 3, these data are altitude corrected by multiplying reported velocities by 0.86, the average barometric pressure (in bars) at that altitude.

4.   The use of the local average velocity, V, appears to adequately correlate the effect of variations in dimensionless channel height, H/a, and packaging density, D.

Observations 2 and 4 in the above list suggest that the appropriate length scale to unify the data is the package length, L. Figure 4 shows the data rendered in these terms where the package Nusselt number, $Nu_L = hL/k$ and package Reynolds number, $Re_L = \rho VL/\mu$, are both based on the package length. The bandwidth of the data is now about ±10% Furthermore, the figure shows a change in the slope of the Nusselt–Reynolds relation suggestive of laminar–turbulent transition at $Re_L \approx 5000$. Power-law correlations for the data give

$$Nu_L = 0.6\,Re_L^{0.5}\,Pr^{0.33}, \quad Re_L \leq 5000 \tag{9}$$

and

$$Nu_L = 0.082\,Re_L^{0.72}, \quad Re_L > 5000 \tag{10}$$

where Pr is the coolant Prandtl number. Equation 9, for low Reynolds number flow, is seen to be a simple 10% downward adjustment of the Blasius solution

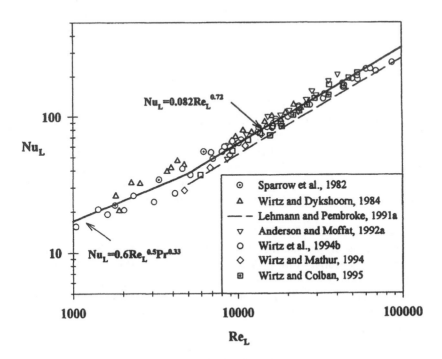

FIGURE 4. Correlation of data from measurements listed in Table 1 for package Nusselt number in terms of package Reynolds number.

for laminar flow over an isothermal plate of length, L. Equation 10, for the higher Reynolds numbers, is fully empirical. It gives heat transfer coefficients that are 40 to 75% higher than would be obtained using the Stanton number correlation for turbulent flow over a smooth, isothermal plate [Kays and Crawford, 1993]. In both cases, the data is seen to scatter $\pm 10\%$ about the correlations. Since 5–10% (in both Nu and Re) is generally the limit of accuracy for the experimental results considered, further refinement such as inclusion of correction factors for variations in packaging density, package height, or channel height, seems unnecessary.

For low profile packages having small but finite package heights, the correlation of Lehmann and Pembroke [1991a], shown as a dashed line in the figure, is recommended[5]

$$Nu_L = 0.07\, Re_L^{0.718} \qquad (11)$$

Other correlations for the heat transfer coefficient have been suggested. For example, Anderson and Moffat [1992b] developed a linear correlation of the heat transfer coefficient in terms of the coolant turbulence level

$$h = c_1 + c_2 u'_{max} \qquad (12)$$

where u'_{max} is the theoretical maximum turbulence level in the shear layer that separates the bypass-flow from the array-flow. In Equation (12), $c_1 \approx 16$, $c_2 \approx 30$ are empirically determined constants. The quantity u'_{max} is estimated as

$$u'_{max} = 0.82\left[\left(-\frac{1}{\rho}\frac{dp}{dx}\right)\frac{V_o S_x}{L_x}\right]^{\frac{1}{3}} \qquad (13)$$

where $-dp/dx$ is the coolant flow pressure gradient. Streamwise turbulence measurements by Wirtz and Chen [1992] in a channel containing two-dimensional ribs having an aspect ratio of 6 show that h(2-D) is nearly linearly proportional to the measured maximum streamwise turbulence level, u'_{max} (max). Unfortunately, correlation of the data of Table 1 in terms of u'_{max} shows that c_1 and c_2 in Equation 12 are functions of package length, L, and array density, D. Equation 12 over-predicts measured values of h for packages having L > 50 mm by 20–40%.

THERMAL WAKE FUNCTION

Correlations of the thermal wake function, θ_{k-i}, have been proposed by Arvizu and Moffat [1982], Wirtz and Dykshoorn [1984], Anderson and Moffat [1992], and Kang [1994]. Kang considers a point source of heat in a turbulent channel flow and finds that θ_1, the wake function for an array element immediately downstream from the heat releasing element, is given by the following expressions[6]:

---

[5]   Lehmann and Pembroke also present correlations for low Reynolds number flows.

[6]   Kang also gives expressions for the thermal wake function in laminar flow. These expressions are not presented here since experimental verification of these expressions is not available.

$$\theta_1 = 7.19C \, \text{Pr}^{-0.5} \left(\frac{L_x}{S_x}\right)^{0.5} \left(\frac{L_x}{H}\right)^{0.44} \left(1 - \frac{a}{H}\right)^{0.5} \text{Re}_{o,L}^{n-0.94} \qquad (14)$$

where $\text{Re}_{o,L} = V_o \text{Re}_L / V$ is the package Reynolds number based on inlet average velocity, V_o. The constant, C, and exponent, n, in Equation 14 are the coefficient and exponent, respectively, of the power-law that correlates the heat transfer coefficient for the packages

$$Nu_L = C \, \text{Re}_{o,L}^n \qquad (15)$$

The thermal wake function for successive packages, $\theta_{k-i} \, k > i + 1$, is given in terms of θ_1 by the expression

$$\theta_{k-i} = \theta_1 \left(\frac{1}{k-i}\right)^m \qquad (16)$$

where the exponent m has been experimentally determined to range from 0.5 to 1.0. Kang shows that m is a function of the turbulent Peclet number of the flow. A curve fit of his numerical results gives

$$m \cong .5 + 0.06 e^{-\frac{Pe_t}{2500}} \qquad (17)$$

where $Pe_t = VL_y^2 / \alpha_t S_x$ is the turbulent Peclet number and $\alpha_t = 0.006 \, \alpha (VH/\nu)^{.88}$ is an estimate of the turbulent thermal diffusivity. The electronics cooling applications considered here have Pe_t ranging from about 250 to 1000, so $m \cong 0.55 \pm 0.01$.

Table 2 compares estimates of the thermal wake function given by Equations 14 and 15 with experimentally determined correlations of Wirtz and Dykshoorn [1984] for two dimensionless channel wall-to-wall spacings. In developing these expressions, Equation 10 has been used for the package heat transfer coefficient correlation, thus the coefficient, C, in Equation 15, is given as $C = 0.082(V/V_o)^{0.72}$, where V/V_o is given by Equation 2. The results show that the fractional temperature increase of the element immediately downstream from the heat-releasing element decreases with increasing flow rate. The theoretical result predicts a weaker Reynolds number dependence than was found in the experiments. However, as shown in Figure 5, the agreement between the theoretical result and the empirical correlation is very good in both cases, although it is anticipated

TABLE 2

**Comparison of Thermal Wake Function Estimates of Kang (1994)
with Data Correlations of Wirtz and Dykshoorn (1984)**

Investigator	H/a = 2	H/a = 5
Kang (1994), theoretical	$\theta_1 = 0.58 \, \text{Re}_{o,L}^{-.22}$	$\theta_1 = 0.43 \, \text{Re}_{o,L}^{-.22}$
Wirtz and Dykshoorn (1984), experiments	$\theta_1 = 1.29 \, \text{Re}_{o,L}^{-.3}$	$\theta_1 = 0.82 \, \text{Re}_{o,L}^{-.3}$

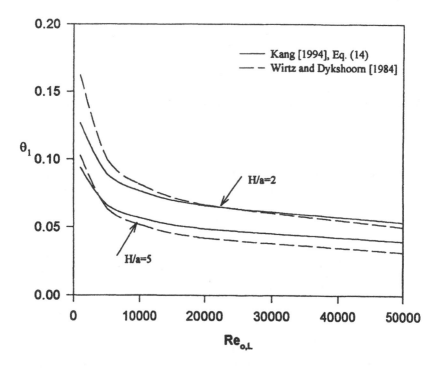

FIGURE 5. Comparison of Equation 14 for the thermal wake function with empirical correlation of Wirtz and Dykshoorn [1984].

that Equation 14 will overpredict θ_1 at high Reynolds numbers and high dimensionless channel heights. The figure shows that an increase in channel wall-to-wall spacing reduces the thermal wake effect. Furthermore, it is noted that heat release from an element in row "i" of an array results in a 5–10% fractional temperature increase for the element immediately down stream in row (i + 1).

Figure 6 compares the predictions of Equation 16 and 17 with the data of Wirtz and Dykshoorn [1984] for the temperature rise of successive elements further downstream from the heated element. The case shown is for two Reynolds numbers with H/a = 2. For the higher Reynolds number, Pe = 375 and Equation 17 gives m = 0.55. This results in a good fit of the data, as can be seen in the upper part of the figure. At the lower Reynolds number ($Re_{o,L}$ = 1355), Pe = 285 and Equation 17 gives m ≅ 0.55. In this case, the data are better fit if m = 0.78, and the theoretical result is seen to overpredict the wake function by about 20–30%, as shown in the lower part of the figure.

EXAMPLE CALCULATION

We can gain an appreciation of the relative importance of these effects (self heating and thermal wake) by considering an example. Consider an array of square planform packages having package length, L = 36 mm, and height, a =

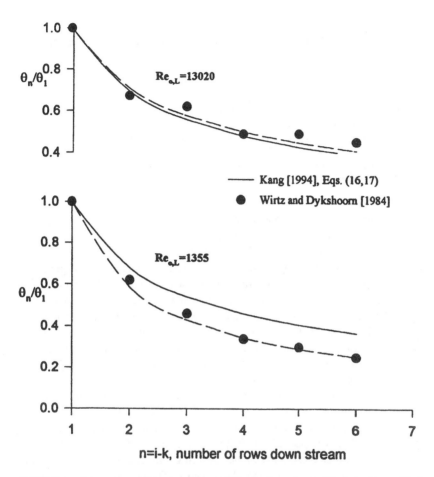

FIGURE 6. Comparison of Equations 16 and 17 for the thermal wake function with empirical correlation of Wirtz and Dykshoorn [1984].

5 mm. The array is five rows long. The packages are spaced c = 8 mm apart (D = 0.67), and the board-to-board spacing is H = 20 mm. Air at 20°C flows toward the array at 800 fpm (V_o = 4.1m/s). Under these conditions V/V_o = 1.26, the channel Reynolds number is Re = 13,600, and the package Reynolds number is Re_L = 12,300. Equation 10 gives a fully developed flow heat transfer coefficient, h = 52W/m²°C. Equations 14 through 17 give θ_k, k = 1, . . , 4 as (0.110, 0.075, 0.060, 0.051). Table 3 summarizes the resulting operating temperatures, $T_k - T_o$, when the first row, or all rows, are powered at 5W per element.

The first row of the Table shows package temperature increases (relative to the entrance flow temperature) for the case where only the element in the first row is powered and no correction is made for entrance effects. Self heating results in a 47.7°C temperature increase of the first package, and the temperature increase of successive (unpowered) elements is relatively small.

TABLE 3
Estimated Package Temperature Increase Due to Convection Only

Row number, r	Package Temperature Increase, $T_r - T_o$, °C				
	1	2	3	4	5
only q_1 = 5W, no entrance correction	47.7	5.3	3.6	2.9	2.5
all q's = 5W, no entrance correction	47.7	52.9	56.5	59.4	61.8
all q's = 5W, with entrance correction	39.7	48.6	56.5	59.4	61.8

Note: Cooling conditions are defined in the accompanying text. Temperature increases are rounded
to one digit of significance.

The second row of the table demonstrates how these seemingly small thermal wake effects are compounded by multiple heat sources. The case considered is for all rows powered at 5W per element with no correction for entrance effects. Successive element temperature increases are obtained via Equation 4 by adding the thermal wake temperature increase contribution of each upstream heat-releasing element in the same array column to the self-heated temperature rise of the element under consideration. For example, the temperature increase of the third-row element is 47.7 + 5.3 + 3.6 = 56.6 (intermediate calculations rounded to 1-digit). In this case, heat release from the first four rows results in a 30% increase over the selfheated temperature rise in the surface temperature of the fifth-row element.

Finally, the third row of the Table demonstrates how entrance effects affect package operating temperatures. In this case, we assume the first and second row heat transfer coefficient is augmented by 20% and 10%, respectively, relative to the fully developed value of h. The results show 17% and 8% reductions relative to uncorrected estimates in the operating temperatures of the first- and second-row elements. Furthermore, the entrance effect does not extrapolate to successive element rows past the first two rows. Since the heat release rate from the first and second row elements is fixed (at 5W each), the sturcture of the thermal wake downstream from these elements is not changed by the assumed augmentation in the element heat transfer coefficient.

COOLANT PRESSURE DROP—
UNIFORM, IN-LINE ARRAYS

Upon encountering a package array, the coolant is accelerated because of the reduced cross-sectional area available to the flow, and the shape of the velocity distribution changes. As a result, there is an increase in the streamwise pressure gradient, -dp/dx. As the entrance effect diminishes (three to five package rows) the local pressure gradient reduces to a constant (in a row-periodic sense), fully developed value. The fully developed pressure gradient can be expressed in dimensionless terms as a friction factor based on the bypass flow, f, as

$$f = \frac{\left(-\dfrac{dp}{dx}\right)2\,(H - a)}{\dfrac{1}{2}\rho V^2} \qquad (18)$$

Dimensional analysis shows that f is a function of Reynolds number and the geometric lengths that describe the array. Figure 7 shows some typical data [Wirtz et al., 1994b, Wirtz and Colban, 1995] where the friction factor is plotted in terms of the channel Reynolds number, Re. The upper part of the figure compares two arrays having different packaging densities, and the lower part compares two channel heights.[7] Also shown in the figure are laminar and low-Reynolds number turbulent [Beavers et al, 1971] correlations for flow in a smooth-wall channel having a height equal to (H-a). The smooth-wall correla-

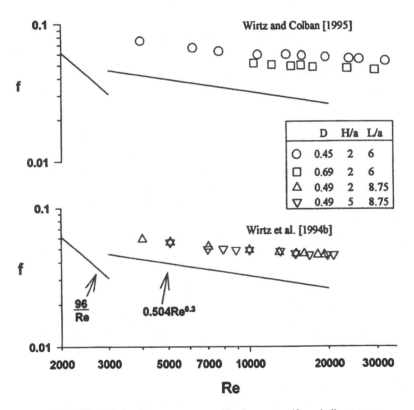

FIGURE 7. Friction factor measurements for flow over uniform, in-line arrays.

[7] Since Re is defined in terms of local average velocity, V, a comparison of f for different geometrical configurations does not imply a comparison at equal entrance flow rates.

tions can be interpreted as applicable to arrays having maximum packaging density (D → 1).

The data exhibit turbulent-like behavior, with the friction factor slowly decreasing with increasing Re. The slope of the plot of data in log–log coordinates suggests $f \sim Re^{-0.2}$. This implies that the pressure drop across an array, Δp, is proportional to $V_o^{1.8}$ so that a doubling in flow rate results in a factor-3.5 increase in Δp. The upper part of the figure shows that f decreases with an increase in packaging density, presumably approaching the smooth-wall correlation as D increases. The lower part of the figure shows that the present friction factor definition, Equation 18, appears to adequately correlate variations in channel height. As a result, for a given array we expect $\Delta p \sim (H-a)^{-1}$. Furthermore, a comparison of the two plots suggests that f decreases with increase in the package aspect ratio, L/a. It must be pointed out that there is only a limited pressure-drop data set available, therefore a comprehensive correlation of these effects is not feasible.

NONUNIFORM ARRAYS

EFFECT OF ARRAY STAGGER

The effect of staggering alternate rows of packages has been studied by Hollworth and Fuller [1987], Garimella and Eibeck [1990], and Wirtz and Colban [1995]. Bazydola and Taslim [1993] have looked at entrance region effects on a staggered array. At fixed coolant flow rate, stagger increases both the package heat transfer coefficient (by up to 50%) and the friction factor (by up to a factor of two). However, Wirtz and Colban found that there is no cooling performance advantage resulting from row stagger on the basis of equal coolant pumping power such as exists in a fan-driven system. Furthermore, if cooling performance is expressed in terms of heat removal per unit system volume, maximum capacity is expected with the most dense packaging possible.

EFFECT OF NONUNIFORM PACKAGE SIZE

Sparrow et al. [1984], and Torikoshi et al. [1988] considered how changes in array element height affected the package heat transfer coefficient. In general, they found that package heat transfer is slightly enhanced in the vicinity of the change in geometry.

EFFECT OF HEAT SINK PLACEMENT

Lehmann and Kosteva [1990], Lehmann and Pembroke [1991b], and Gavali et al. [1993] studied how placement of a heat sink on an array element affects the thermal performance of other array elements in the array. In general, elements immediately adjacent to the heat-sinked element (in the adjacent columns) experience a small amount of augmentation in their heat transfer coefficient (about 5–10%) due to local acceleration of the coolant around the blockage caused by the heat sink. Array elements downstream from the heat-sinked element may experience either an increase or a decrease in their heat transfer rate due to the heat sink's presence. Wirtz et al. [1994a] explain how this effect influences operating temperatures. Wirtz et al. [1994b] present correlations for computing longitudinal fin heat sink performance in arrays of packages, and Lee [1995] presents a calculation methodology for estimating heat sink performance.

CONCLUSION

Convection in uniform arrays containing low profile packages is strongly influenced by the bypass flow. The primary scaling quantities that control the magnitude of the heat transfer coefficient are the local average velocity, V (an estimate of the bypass velocity), and the package length, L_x. Other geometric parameters that influence h are the packaging density, D, package height, a, and the channel height, H. Equations 9 and 10, heat transfer correlations based on modeling experiments, appear to adequately incorporate these effects in situations where the height of the low profile package is sufficient to modify the coolant flow. These correlations estimate the package heat transfer coefficient with an uncertainty of about ±10%. If the package height is very small, Equation 11 is recommended. In a similar manner, Equations 14–17 allow for estimation of the thermal wake function. These two effects, self heating and thermal wake, may then be superimposed using Equation 4 to estimate package surface temperatures under conditions where radiation and conduction are neglected.

APPENDIX A: NOMENCLATURE

a	Package height
A	Package surface area
c	Package spacing
D	Packaging density
f	Friction factor
h	Adiabatic heat transfer coefficient
H	Board-to-board spacing
k	Thermal conductivity
L	Package length
Nu	Nusselt number
p	Pressure
q	Convective heat transfer rate
R_{c-a}	Case-to-ambient thermal resistance
Re	Channel Reynolds number, $V2(H-a)/v$
Re_L	Package Reynolds number, VL/v

S   Packaging pitch
T   Temperature
V   Local average velocity
V_o Inlet average velocity
x   Streamwise coordinate
y   Cross-stream coordinate
z   Spanwise coordinate
α Thermal diffusivity
μ Viscosity
ν Kinematic viscosity
θ_{k-i} Thermal wake function, $(T_k-T_o)/(T_i-T_o)$
ρ Density

REFERENCES

Anderson, A.M. and Moffat, R.J. 1992a. The adiabatic heat transfer coefficient and the superposition kernel function: Part I—Data for arrays of flatpacks for different flow conditions. *J. Electron. Packag.*, 12:14–21.

Anderson, A.M. and Moffat, R.J. 1992b. The adiabatic heat transfer coefficient and the superposition kernel function: Part II—Modeling flatpack data as a function of channel turbulence, *J. Electron. Packag.*, 112:14–21.

Arvizu, D.E. and Moffat, R.J. 1982. The use of superposition in calculating cooling requirements for circuit board mounted electronic components, IEEE paper CH1781-4/48-0133.

Beavers, G.S., Sparrow, E.M., and Lloyd, J.R. 1971. Low reynolds number turbulent flow in large aspect ratio rectangular ducts, *ASME J. Basic Eng.*, 285–289.

Bazydola, S.M. and Taslim, M.E. 1993. An experimental investigation of a staggered array of heat sinks in the hydrodynamic and thermal entrance regions of a duct. *J. Electron. Packag.*, 115:106–111.

Copeland, D. 1992. Effects of channel height and planar spacing on air cooling of electronic components. *J. Electron. Packag.*, 114:420–424.

Garimella, S.V. and Eibeck, P.A. 1990. Heat transfer characteristics of an array of protruding elements in single phase forced convection, *Int. J. Heat Mass Transfer*, 33(12): 2659–2669.

Garimella, S.V. and Eibeck, P.A. 1992. Onset of transition in the flow over a three-dimensional array of rectangular obstacles, *J. Electron. Packag.*, 114:251–255.

Gavali, S., Kailish, C., Patankar, S.V., and Miura, K. 1993. Effect of heat sink on forced convection cooling of electronic components, *Adv. in Electron. Packag.*, 1993, ASME EEP, 4(2):801–808.

Hollworth, B.R. and Fuller, H.A. 1987. Heat transfer and pressure drop in a staggered array of air cooled components, in *Proc. Int. Symp. on Cooling Technol. for Electron. Equipment*, Honolulu, 732–748.

Kang, S.S. 1994. The thermal wake function for rectangular electronic modules, *J. Electron. Packag.*, 116:55–59.

Kays, W.M. and Crawford, M.E. 1993. *Convective Heat and Mass Transfer*, 3rd. Ed., McGraw-Hill, New York.

Lee, S. 1995. Optimum design and selection of heat sinks, in *Proc. 11th IEEE SEMI-THERM Symposium*, San Jose, 48–54.

Lehmann, G.L. and Kostiva, S.J. 1990. A study of forced convection direct air cooling in the downstream vicinity of heat sinks, *J. Electron. Packag.*, 112:234–240.

Lehmann, G.L. and Pembroke, J. 1991a. Forced convection air cooling of simulated low profile electronic components. I. Base case, *J. Electron. Packag.*, 113:21–26.

Lehmann, G.L. and Pembroke, J. 1991b. Forced convection air cooling of simulated low profile electronic components. II. Heat sink effects, *J. Electron. Packag.*, 113:27–32.

Moffat, R.J. and Anderson, A.M. 1990. Applying heat transfer data to electronics cooling, *J. Heat Transfer*, 112:882–890.

Sparrow, E.M., Niethammer, J.E., and Chaboki, A. 1982. Heat transfer and pressure drop characteristics of arrays of rectangular modules encountered in electronic equipment, *Int. J. Heat Mass Transfer*, 25(7):961–973.

Sparrow, E.M., Yanezmoreno, A.A., and Otis, D.R. 1984. Convective heat transfer response to height differences in an array of block-like electronic components, *Int. J. Heat Mass Transfer*, 27:469–473.

Torikoshi, K., Kawaxoe, M., and Kurihara, T. 1988. Convective heat transfer characteristics of arrays of rectangular blocks affixed to one wall of a channel, *Natural and Mixed Convection in Electronic Equipment Cooling*, ASME HTD, 100:59–66.

Wirtz, R.A. and Chen, W. 1992. Laminar-transitional convection from repeated ribs in a channel flow, *J. Electron. Packag.*, 114:29–34.

Wirtz, R.A., Chen, W., and Colban, D.M. 1994a. Convection in arrays of electronic packages containing longitudinal fin heat sinks, in *Cooling of Electronic Systems*, S. Kakac, H. Yuncu, and K. Hijikata, Eds., Kluwer Academic, Dordrecht, The Netherlands, 145–164.

Wirtz, R.A., Chen, W., and Zhou, R. 1994b. Effect of flow bypass on the performance of longitudinal fin heat sinks, *J. Electron. Packag.*, 116:206–211.

Wirtz, R.A. and Colban, D.M. 1995. Comparison of the cooling performance of staggered and in-line arrays of electronic packages, in *Proc. ASME/JSME Thermophys. Conf.*, Maui, 4:215–221.

Wirtz, R.A. and Dykshoorn, P. 1984. Heat transfer from arrays of flat packs in a channel flow, in *Proc. 4th IEPS conf.*, Baltimore, 318–326.

Wirtz, R.A. and Mathur, A. 1994. Convective heat transfer distribution on the surface of an electronic package. *J. Electron. Packag.*, 116:49–54.

FOR FURTHER INFORMATION

Basic information on thermal control in electronics is found in *Thermal Analysis and Control of Electronic Equipment*, A.D. Kraus and A. Bar-Cohen (McGraw Hill, 1983). More advanced summaries on heat transfer in electronics are found in *Advances in Thermal Modeling of Electronic Components and Systems, Vols. 1, 2 and 3*, A. Bar-Cohen and A.D. Kraus, Eds. (ASME Press).

The Electrical and Electronics Packaging Division (EEPD) of ASME acts as a coordinator and disseminator of information on electronic packaging. The division's *Journal of Electronic Packaging* contains technical papers on recent developments in heat transfer in electronics. The K-16 Committee on Heat Transfer in Electronics (a committee of EEPD and the Heat Transfer Division of ASME) acts as a forum for exchange of technical information.

Chapter 4

CONJUGATE HEAT TRANSFER IN FORCED AIR COOLING OF ELECTRONIC COMPONENTS

Alfonso Ortega

CONTENTS

0-8493-9447-3/96/$0.00+$.50
© 1996 by CRC Press, Inc.

INTRODUCTION

In air cooling of electronics, the heat transfer process from single chip packages mounted on circuit packs or boards is a combined mode process involving conduction within the chip carrier and the board and convection from the package and board surfaces. There are numerous approaches for analyzing the thermal behavior of the package/board system including traditional uncoupled approaches that apply convective heat transfer coefficient boundary conditions to the solid conducting domains and more rigorous coupled approaches that acknowledge the conjugate nature of the heat transfer process through the multiple paths from the chip to the package to the board and eventually to the convective air flow. In assessing modeling strategies for electronics thermal management, the need for several approaches, ranked in a hierarchy according to application, ease of use, computational requirements, and accuracy, is frequently overlooked. A traditional uncoupled conduction analysis of a packaging structure, such as a board with convective heat transfer coefficients imposed as surface boundary conditions, is approximate at best, but it has a low computational overhead, is relatively easy to develop, and may lead to satisfactory results for a system analysis. On the other hand, a full conjugate approach requiring simultaneous solution of equations describing diffusion of heat in the solid and diffusion and advection of heat in the fluid may carry a higher computational penalty but will generally be more accurate when the fluid flow can be modeled well. As computational power has become increasingly available to the thermal analyst, the ability to perform reasonably accurate flow simulations coupled directly to the thermal analysis has rapidly evolved.

The present chapter is devoted to a description of some selected conjugate problems in air cooling of electronic components. There are several objectives. The first is to introduce nonspecialists to the subject by discussion of key features of conjugate problems involving electronic components on boards. It is hoped that the discussion will provide a starting point for understanding the differences between coupled and noncoupled approaches and in particular point out difficulties in the use of the heat transfer coefficient when conjugate conduction mechanisms are important. For specialists, the aim of the chapter is to illustrate the use of various approaches for conjugate analysis and to present physical insight that has recently been gained by their application in the author's research group. The experienced analyst may wish to skip the background material as it is intended primarily to engage the less experienced reader who may be encountering the subject for the first time.

BACKGROUND

In principle, convective heat transfer from a solid surface to a flow is always a conjugate process involving the combined and simultaneous action of conduction in the solid wall and advection/diffusion in the fluid. In other words, all practical heat transfer problems, including those that arise in air cooling of electronics, are conjugate in nature. The fact that the word "conjugate" is so frequently associated with the electronics cooling problem, when in fact all practical convective heat transfer problems are conjugate, tells more about the historical evolution of convective heat transfer than the nature of the electronics cooling problem. The electronic package structures examined in this chapter are illustrated in Figure 1, where it is understood that many single chip components may reside on a single circuit pack. What is so special about this problem, and why does it carry the "conjugate" descriptor? To elucidate, examine the two problems depicted in Figure 2, each with identical flow over upper and lower board surfaces, with boards of identical thickness but with different board thermal conductivity and heat dissipation conditions on the surfaces. In Figure 2(a) the surfaces of the board are entirely given up to uniformly dissipating sources of identical strength, $\dot{q}$. Because the board is non-conducting, the board has no effect on the temperature distribution and the surfaces may just as well be represented with a uniform heat flux thermal boundary condition, i.e.

$$-k_f\left\{\frac{\partial T}{\partial y}\right\}_{y=0} = \dot{q}, 0 \leq x \leq L \tag{1}$$

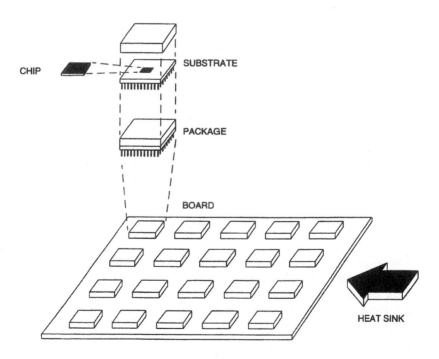

FIGURE 1. A hierarchical representation of chip, chip carrier, and board packaging levels.

where k_f is the fluid thermal conductivity, T is fluid temperature, and $\dot{q}$ is the heat dissipation per unit area. Contrast the just described situation to that of Figure 2(b) in which a single source with total heat dissipation $\dot{q}_{total}$ is located on a board of finite thickness and thermal conductivity. In this case, the heat dissipated by the source may be conducted into the board and released from either surface, resulting in a nonuniform distribution of heat flux from the board to the fluid on the board surfaces. To the thermal engineer's distress, the heat flux distribution can no longer be specified on the board surfaces *a priori* as in the first problem because it is unknown. Equally disturbing is the fact that the distribution of convective heat transfer coefficient, so nicely catalogued in countless heat transfer textbooks for cases such as in Figure 2(a), is completely different for case 2(b), despite the fact that the flow field is identical.

One approach to the thermal problem is to adopt an uncoupled strategy in which all information about the convective part of the problem is embodied in a convective conductance, or heat transfer coefficient, h, specified on the solid surfaces. For the problem of Figure 2(b), the Laplace equation on the board may be solved using any of a number of approaches, most commonly by applying a finite-difference or finite element approach, subject to convective heat transfer coefficient boundary conditions on the board surface. The art of choosing a correct representation for the convective heat transfer coefficient h on the board surfaces will be the key to a successful model. In the application at hand, the local heat transfer coefficient is sometimes defined in terms of the undisturbed fluid temperature as

$$h = \frac{\dot{q}}{T_s - T_o} \tag{2}$$

where T_s is the local surface temperature and T_o is the air temperature far from the plate surface.

Here, $\dot{q}$ is interpreted as the local heat flux entering the fluid from the surface. A typical solution, for example, for laminar boundary layer flow over a smooth surface with conditions of constant heat flux on the wall, such as in Figure 2a., is commonly found in heat transfer monographs. For example, Kays and Crawford [1993] give the solution as

$$Nu_x = \frac{hx}{k_f} = 0.453 Re_x^{1/2} Pr^{1/3} \tag{3}$$

where Nu_x is local Nusselt number, x is local streamwise position, Re_x is local Reynolds number, $Re_x = Ux/v$, and Pr is fluid Prandtl number. It is seen as in Figure 2(c) that h declines as $x^{-1/2}$ on the surface, a well-behaved monotonically decreasing behavior, and that it gets infinitely large at $x = 0$. This latter behavior is unrealistic and simply reflects the failure of the boundary layer approximations at small x. In the theory of heat exchangers, which forms the basis for so much of the material found in heat transfer textbooks, these types of boundary conditions are frequently used with great success on convective surfaces despite the fact that they may be uniform neither in temperature nor in heat flux. In general, only the surface averaged heat transfer coefficient is of interest. The effects of wall conduction, especially in the streamwise direction, are generally unimportant in such problems, and thus it is frequently forgotten that solutions such as Equation 3 are not only dependent on the geometry under consideration but also on the thermal boundary conditions and the temperature used as the local reference temperature, in this case, T_o. It would be erroneous to specify a heat transfer coefficient derived from Equation 3 to the surfaces of the board depicted in Figure 2(b) for the purpose of calculating the surface temperature. To illustrate, when problem 2(b) is solved using a true coupled conjugate approach, which did not require *a priori* specification of *h* on the surfaces, the resulting distribution of heat transfer coefficient defined by Equation 2 would appear as in Figure 2(d). In this case, the board conduction produced a highly nonuniform heat flux distribution on the surface that is dependent on the thickness and conductivity of the board. The distribution of *h* reflects the nonuniformity in the distribution of the heat flux. The heat transfer coefficient *h* is badly behaved and is so dependent on the conjugate board conduction that it loses its primary value, which is its generality, and hence is not very useful for predicting the board and component temperatures. Proper solution of a conjugate problem requires an approach that simultaneously considers fluid convection and substrate conduction.

To illustrate conjugate, combined mode heat transfer in an actual air-cooled board, consider the temperatures that result from a single PQFP located on a test board in a low speed parallel air flow. Two such test boards are illustrated in Figure 3. The two boards are at this writing under evaluation by the JEDEC JC15.1 subcommittee on thermal measurements [JEDEC, 1994] with the aim of establishing industry standard procedures and hardware for thermal characterization of single chip packages. The traces on the board of Figure 3(a) are designed to accommodate a single 64 lead PQFP and are not nested. The lead out is conventional and "radial" in design. The board of Figure 3(b) is designed to accommodate packages of different size footprints, and this is accomplished with nested traces that emanate in the four principle axis directions on the board surface. In both cases the copper traces are 2.5 mil thick. The copper traces produce a highly anisotropic thermal conductivity in the boards, with in-plane conductivity much higher than out-of-plane conductivity and with significantly different spatial distribution. Figure 4 presents a series of infrared thermographs made at three air flow speeds for the two boards. The thermographs for the unnested "radial" trace pattern, Figure 4(a) show that the traces conduct heat radially from the package, more or less equally, and that the convection to the air distorts the radial symmetry because of the thermal wake effect. The results for the nested trace pattern, Figure 4(b), were surely anticipated—the trace spatial density is highest in the direction of the principal axes and the resulting preferential conduction via these routes is apparent. The convective wake that develops due to the spatial evolution of the air-side

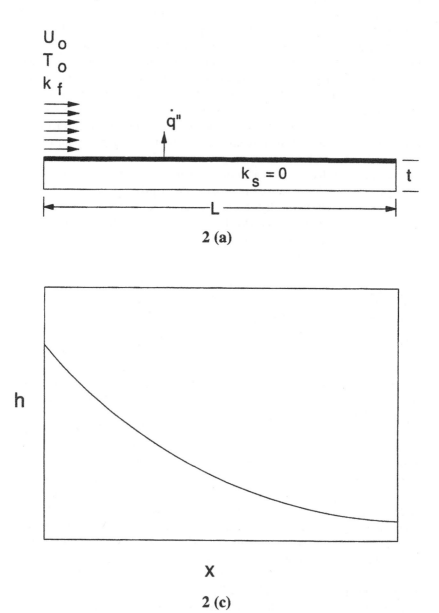

FIGURE 2. Comparison of a nonconjugate and conjugate problem in air cooling of a PCB; (a) board with zero conductivity and uniform dissipation of heat on its surface, (b) board with nonzero conductivity and localized dissipation of heat on its surface, (c) local heat transfer coefficient distribution for case (2a), the nonconjugate problem (d) local heat transfer coefficient distribution for case (2b), the conjugate problem.

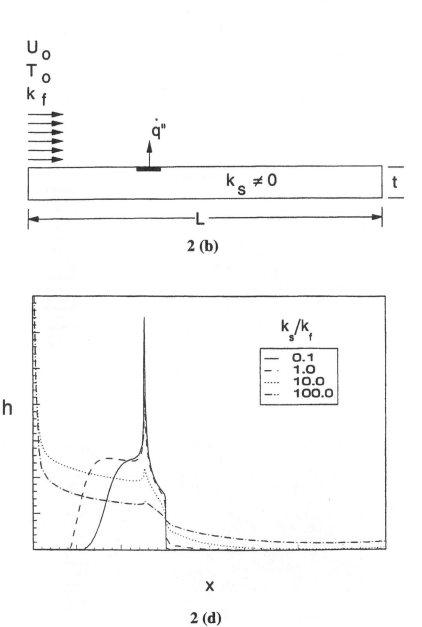

2 (b)

2 (d)

FIGURE 2. Continued.

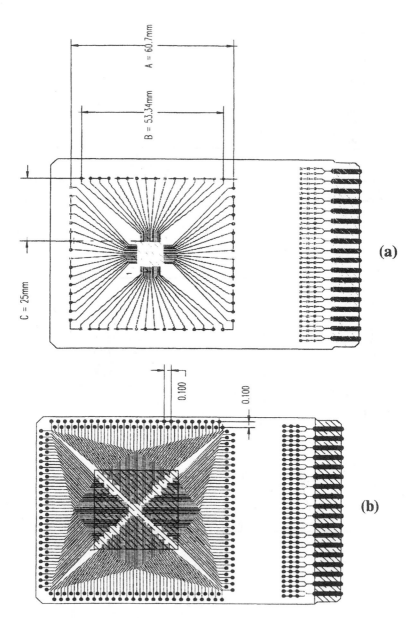

FIGURE 3.   PCB test coupons for Θ_{j-a} measurements of a single Motorola™ 64 lead PQFP in low speed wind tunnel; (a) nonnested copper traces (2.8 mil Cu), (b) nested copper traces (2.8 mil Cu). [Courtesy of Darvin Edwards, Texas Instruments].

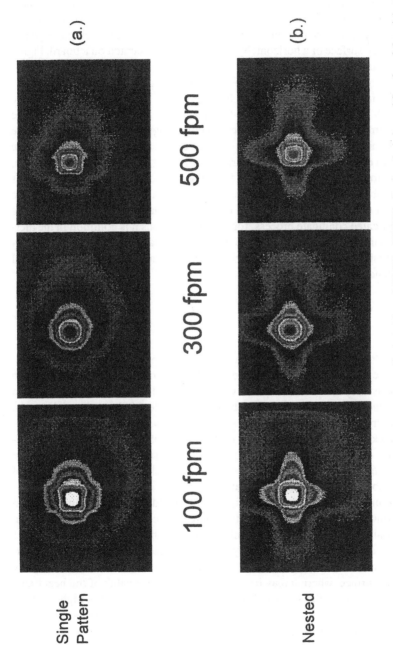

FIGURE 4. Infrared thermographs of the surface temperature of a single Motorola™ 64 lead PQFP in low-speed wind tunnel flow from left to right on (a) nonnested PCB, and (b) nested PCB [Courtesy of Darvin Edwards, Texas Instruments].

thermal boundary layer is still apparent in the flow direction. In both cases, the distribution of heat flux on the surface of the package and the board is highly nonuniform and is dependent on the trace pattern and board thickness and conductivity. A second illustration of conduction and convection acting simultaneously is shown in Figure 5, which shows two infrared photographs of the top surface of a horizontally oriented PQFP mounted on a board. Figure 5(a) is an infrared thermograph of the top surface of the PQFP cooled by natural convection to air, and Figure 5(b) is the same package cooled by air flowing from left to right at 1 m/s. The heat spreading by conduction from the die through the plastic to the PQFP top surface is readily observed from the isotherms. In the case of natural convection, Figure 5(a), the isotherms are nearly circular, indicating that the surface convective resistance to heat flow is more or less uniform over the surface, or at least circumferentially uniform. In the forced cooling case, Figure 5(b), the convective thermal wake effect is again observable by the elongation of isotherms in the flow direction. Despite the fact that the heat transfer from the heat-dissipating die to the air flow is a complicated conjugate process, for practical reasons it is common in the thermal characterization of single chip packages to characterize the package thermal resistance with simple metrics such as junction to ambient thermal resistance, Θ_{j-a}, defined as

$$\Theta_{j-a} = \frac{T_j - T_{ambient}}{P_d} \tag{4}$$

where P_d is total power dissipation from the die and T_j is the average die temperature. It is not surprising that the use of these simple metrics is fraught with inconsistency and misconception. One reason is the general unwillingness to reject the notion that surface heat transfer coefficient h depends solely on the geometry and the flow descriptors such as velocity. There appears to be a general lack of appreciation of the fact that in most problems in air cooling of electronics, h depends simultaneously on the characteristics of the flow *and* on the characteristics of the conducting solids of the package structure, i.e., most of the problems are "conjugate" in nature.

Returning to the example of Figure 2(b), the degree of coupling between the convective heat transfer from the source and conduction in the board can be parameterized by the Biot number based on the board thickness. Table 1 compares the thermal conductivity and the inverse Biot number ($Bi^{-1} = k/ht$) of a 1-mm-thick layer of various packaging materials with a convectively cooled surface, where h may be a nominal or average value of the heat transfer coefficient on the surface. If the heat transfer coefficient is high, or if the substrate conductivity is low, coupled conduction effects in the substrate are negligible, as, for example, is the case for a substrate composed of balsa wood with Bi^{-1} order 1.0. Conversely, as substrate thermal conductivity increases, or if the convective heat transfer from the surface is low, conjugate conduction

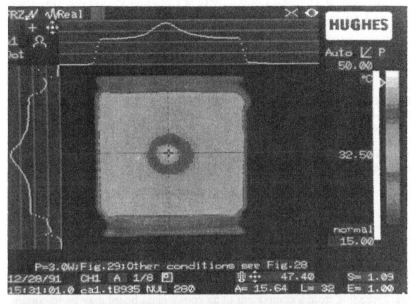

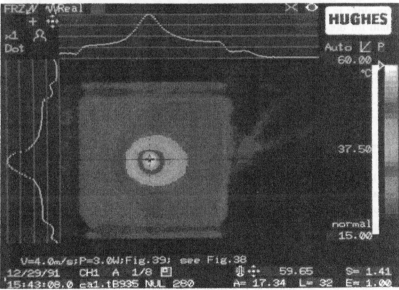

FIGURE 5. Infrared thermographs of the upper surface of a horizontally oriented PQFP under conditions of natural convection air cooling (top), and forced convection air cooling at a velocity of 1 m/s (bottom). [Courtesy of Prof. Zhu Dezhong, Mechanical Engineering Department, Tsinhua University, Beijing, China].

TABLE 1

Comparison of Thermal Conductivity of Common Packaging Materials and Their Inverse Biot Number for a Typical Comparison Case

Material	Thermal Conductivity k (W/m-K)	Biot Number^{-1}(k/ht) h = 50 W/m^2-K t = 1 mm
Balsa Wood	0.05	1.0
Epoxy—Glass	0.18 (out of plane)	3.6
Plastic Molding Compound[1]	0.7	14.0
Epoxy—Glass	1.1 (in plane)	21.7
Alumina	19.7	384.6
Al-Nitride	170	3333.3

Source: Properties abstracted from Jensen et al. [1989], except for [1]Tummala and Rymaszewski [1989], with permission.

mechanisms may be important. Thus, from Table 1, lateral conduction and heat spreading from a component to an epoxy-glass board may be important to consider in a low speed, air-cooled application and will certainly be important for the ceramic substrates. Although the length scale used in Table 1 is somewhat arbitrary, the point to be made is that the magnitude of Bi^{-1} is indicative of the strength of the coupling between the convection and the solid conduction.

CLASSIFICATION OF PROBLEMS

The class of problems that are of interest here are illustrated in Figure 6, which shows a series of subproblems of the general printed circuit board (PCB) cooling problem [Ortega, et al., 1994]. Starting from the top, the problems are identifiable by geometry, thermal boundary conditions, and type of flow phenomena induced. They also increase in physical feature and computational difficulty but follow a tractable path to the most general practical problem, namely the problem of Figure 6(f). In each case the flow may be further organized according to whether it is an unconstrained boundary layer flow over the surface or a channel flow between the surface and its opposing neighbor board, and whether the flow is laminar, turbulent, or transitional. The situations of Figure 6(a) and 6(b) introduce the discrete, conjugate nature of the heat dissipation in a one-dimensional and two-dimensional sense without any flow complexities that may arise in the more difficult geometries. The situations of Figures 6(c) and 6(d) introduce predominately two-dimensional, separated, unsteady, mean flow both on the rib surfaces and on the board. The heat transfer from the two-dimensional rib-like components may be highly coupled to the board conduction and may require a conjugate treatment. The low magnitude of spanwise velocity components may allow a two-dimensional computation, and results for this class of conjugate problem have been reported by numerous investigators in the laminar regime. The protruding components introduce premature transition and high levels of near-wall turbulence even at low Reynolds numbers, and hence the applicable range of laminar computations is limited.

The extension to three dimensions results in the situations of Figures 6(e) and 6(f). The mean flow in these situations is far more complex, involving secondary flows such as the horseshoe vortex that wraps around the base of the upwind components and three-dimensional, unsteady, reversing flow in the region between components. In addition, the air flow in the array, at entrance velocities ranging from 1 to 4 m/s, is in a highly unstable, intermittent, transitional regime. In the fully turbulent regions, the turbulent motions are extraordinarily high, anisotropic, and obviously not in local equilibrium. Numerous studies have experimentally extracted average heat transfer coefficients on the surfaces of such three-dimensional components under conditions that suppress conjugate mechanisms. Most of the data (see for example Anderson and Moffat [1990a,b]) are for components fabricated from conductive metal such as aluminum or copper, and the components are mounted on thermally insulating boards. Hence, under typical low-speed air cooling conditions, such components are nearly isothermal and board conduction is small. The ensemble of heat transfer coefficients thus measured are weakly dependent on the conduction in the component and in the board and thus depend primarily on geometry and fluid flow parameters. The behavior of the heat transfer coefficient in this type of "uncoupled" problem is discussed in the reviews of Moffat and Ortega [1988] and Peterson and Ortega [1990] among other sources. The present chapter aims to broaden the scope of these previous treatments by pointing out the importance of conjugate phenomena in similar problems.

The remaining discussion will focus primarily on the geometries of Figure 6(a), the two-dimensional strip source of heat on a conducting board, and 6(b), the rectangular source of heat on a conducting board. By limiting the discussion to flows that are well understood, the basic features of conjugate mechanisms can be better observed and generalized and solution techniques can be compared. For completeness and in order to point out the ability to analyze more realistic geometries, a recent investigation of a more specific geometry resembling Figures 6(c) and 6(d) will be briefly discussed. Since there are many possible approaches to the analysis of the conjugate problem, Table 2 lists various types of flows, flow models, and solid models that have been found to be useful in addressing the PCB cooling problem; some but not all of these combinations of flow models and solid models will be discussed.

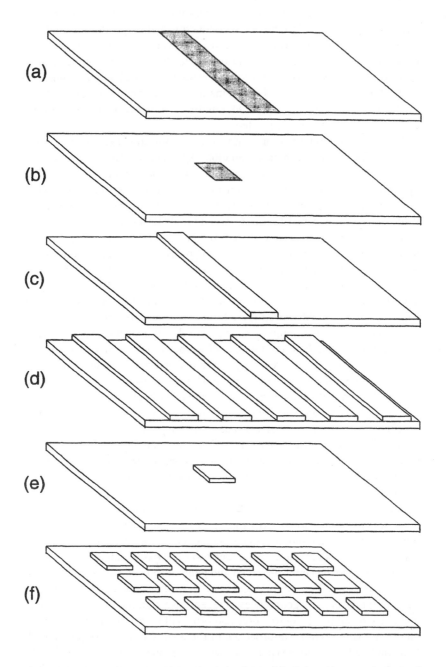

FIGURE 6. Some subproblems of the printed circuit board cooling problem organized according to thermal boundary conditions and geometry [Ortega et al., 1994].

TABLE 2
Types of Flows and Models Useful for Conjugate Analysis of Printed Circuit Boards

Type of Flow	Flow Thermal Model	Solid Thermal Model
Uniform flow	Analytical kernel	Finite element method
Laminar boundary layer flow	Analytical surface conductance	Boundary element method
Laminar channel flow	CFD — finite volume	Finite difference method

TWO-DIMENSIONAL SITUATIONS: STRIP SOURCE OF HEAT IN BOUNDARY LAYER AND CHANNEL FLOW

EXPERIMENTAL OBSERVATIONS

To introduce the relevant physics, results from an experiment specifically designed to examine conjugate heat transfer for the geometry of Figure 6(a) in a relatively simple air flow are first presented [Wirth, 1994; Ortega et al., 1994]. The geometry and thermal conditions of interest are shown in Figure 7, and the experimental realization of the problem is depicted in Figure 8. The apparatus consisted of a low-speed wind tunnel and associated flow conditioning apparatus, a flow development section, and a heat transfer measurement section. Experiments were performed with a single heat transfer section but with various flow development sections. The flow development surface and heat-transfer surface formed one wall of the main flow channel, and the opposite wall was formed by a 0.5 in. plexiglass wall. Smooth plates were used to generate normal flat plate laminar and turbulent boundary layer profiles at the entrance to the heat transfer section. Plates with regular spanwise grooves were used to generate disturbed boundary layers with amplified near-wall turbulence. As shown in Figure 9, the latter plates were grooved such that the rib length B was a constant 1.0 in., the groove depth D was a constant 0.25 in., and the groove length S was 0.25, 0.5, or 1.0 inches.

The heat-transfer surface shown in Figures 8 and 9 was designed to measure conjugate heat transfer from a single 1.0 in. surface-mounted heat source. The panel was 0.5-in.-thick plexiglass (nominal thermal conductivity between 0.18 and 0.2 W/m-K) with a streamwise length of 12.0 in. and a spanwise length of 24.0 in. to match the wind tunnel spanwise dimension. The 1.0-in. heater was 20.0 in. long in the spanwise direction, thus approaching a one-dimensional strip source of heat. It was located 3.0 in. downstream from the beginning of the instrumented heat transfer surface, as shown in Figure 9. The buildup

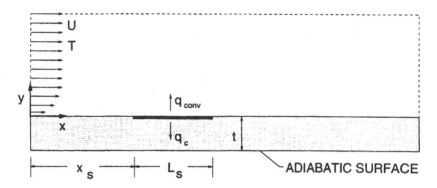

FIGURE 7. Problem domain for a one-dimensional heat source on a conducting substrate [Ortega et al., 1994].

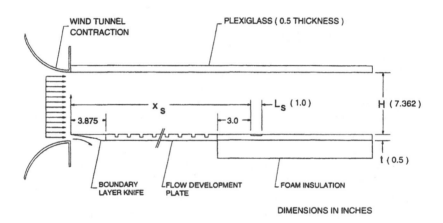

FIGURE 8. Experimental apparatus for measurements of conjugate heat transfer from a one-dimensional heat source on a conducting substrate [Ortega et al., 1994].

of the heater section is shown in Figure 9. The 1.0 in. × 20.0 in. heat source was a 1-mil-thick etched Nichrome foil with a serpentine pattern that provided 95% areal coverage. The foil was encapsulated by 2 mil Kapton film on both sides and was attached to a custom heat flow sensor. The heat flow sensor consisted of a 12-mil-PVDF (trade name KYNAR) spacer with five pairs of evenly spaced 1 mil Type K thermocouples adhered to both sides. Surface temperature measurements were made by embedding 1-mil-diameter Type K thermocouples in a 5-mil-thick adhesive layer adhered to both the flow side and the back side of the 0.5 in. plexiglass panel. The junctions, with diameter of roughly 3 mil, were placed at the spanwise centerline of the heat transfer surface, and the thermocouple wires were laid out along spanwise lines in order to follow isothermal lines on the surface. Relative to the discussion of Table 1, the ther-

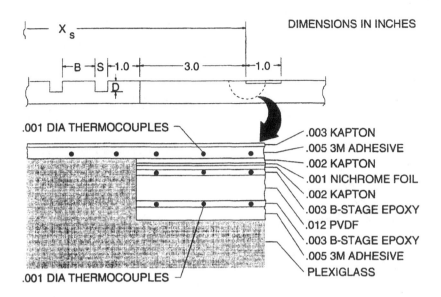

FIGURE 9. Details of development plate and heater section, [Ortega et al., 1994].

mal conductivity of the plexiglass substrate was measured to be 0.19 W/m^2-K, similar to the out-of-plane conductivity of epoxy fiberglass.

Figures 10 and 11 show the mean velocity, the RMS fluctuating velocity, and the percent turbulence level (RMS velocity normalized on mean velocity) for nominally laminar (2 m/s) and turbulent (10 m/s) flows over the smooth plate. The zero pressure gradient Blasius profile is shown for comparison in Figure 10. The laminar flow compares well with the expected theoretical behavior, but the maximum turbulence level of about 1% is rather high and leads to early transition. Similar data for flow developed over the grooved surfaces show significantly increased turbulence levels, up to 55% for a mean flow speed of just 2.0 m/s. The large surface grooves introduce additional modes by which turbulence is generated, especially when they introduce mean flow separation and reattachment. Furthermore, the grooves induce premature transition to turbulence. The presentation of Figure 12, showing the magnitude of the maximum fluctuating velocity as it depends on Reynolds number at a fixed position x, unambiguously shows an inverse monotonic dependence of the transition velocity on groove geometry, as characterized by the length scale B. The most important aspect to be noted is that at the low air velocities commonly encountered in cooling of boards, the flows are invariably unstable transitional flows and the true laminar regime is limited to air velocities of 1 m/s or lower. Furthermore, the transitional velocity is strongly dependent on the surface geometry.

The moderately high plate thermal conductivity and the low rate of convective heat transfer from the surface at these low air velocities combine to pro-

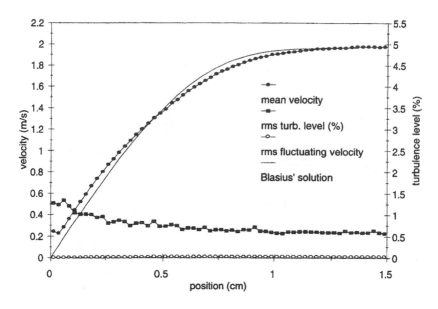

FIGURE 10. Streamwise mean velocity, RMS fluctuating velocity, and % turbulence for flow developed over smooth plate, x = 0.762 m, U = 2 m/s, [Ortega et al., 1994].

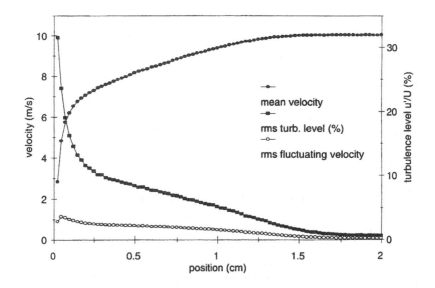

FIGURE 11. Streamwise mean velocity, RMS fluctuating velocity, and % turbulence, for flow developed over smooth plate, x = 0.762 m, U = 10 m/s, [Ortega et al., 1994].

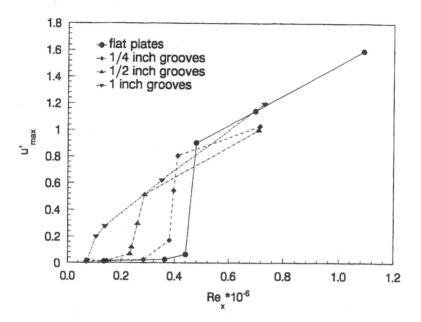

FIGURE 12. Maximum RMS velocity fluctuation dependence on Reynolds number for development plates with increasing groove width, x = 0.762 m, [Ortega et al., 1994].

duce a coupled conjugate heat transfer situation. Defining a nondimensional surface temperature as

$$\Theta = \frac{T - T_o}{\dot{q}_{total} \, L_s/k_f} \tag{5}$$

where $\dot{q}_{total}$ is the total source power dissipation per unit area of source, surface temperatures may be compared under varying conditions of total source power. Figures 13 and 14 show the normalized temperature on the flow side and on the backside, respectively, at freestream speeds varying from 2 to 20 m/s for a smooth development surface. Flow is from left to right, and the heat source is located from position 3 to 4. The effects of the substrate conduction manifest themselves in the significant upwind temperature increase, easily discernible even two source lengths upstream. The temperature peaks on the source downstream from its center and drops steeply downstream in the thermal wake region. The temperature on the rear insulated surface is somewhat more symmetrical because of heat spreading, and the maximum temperature is naturally less than on the heated side. Figure 13 points out the sensitivity of the temperature distribution to transition from laminar to turbulent behavior. There are three cases indicated at a velocity of nominally 10 m/s (Re = 778400); two of the cases, denoted by "weak trip," cluster at a higher temperature, and the

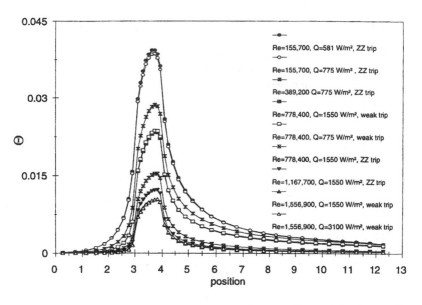

FIGURE 13. Dimensionless surface temperature on flow side with smooth development plate, $x_s = 0.838$ m, [Ortega et al., 1994].

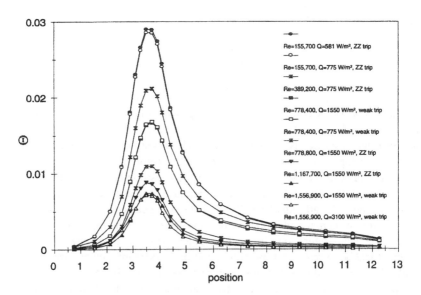

FIGURE 14. Dimensionless surface temperature on back side with smooth development plate, $x_s = 0.838$ m, [Ortega et al., 1994].

third, denoted by "ZZ trip," is significantly lower. The weak trip cases correspond to smooth plates with a single 0.5-in. spanwise strip of sandpaper at the beginning of the smooth plates; this boundary layer trip was not effective in tripping the boundary layer and was replaced by a strip of sandpaper that was placed in a zig-zag pattern on the plate, with a peak-to-peak amplitude of 2.0 in. The ZZ trip was more effective in tripping the boundary layer, and thus, at the same velocity produced lower temperatures.

A typical nonnormalized surface temperature distribution is shown in Figure 15, for a smooth development section and freestream speed of 5 m/s. The 1.0-in. source again is located 3.0 in. from the origin. In the downstream thermal wake region, it is noteworthy that within approximately one heater length scale, the heated side and the back side temperatures become indistinguishable as does conduction in the substrate, implying that the normal temperature gradient in the substrate vanishes. The continued temperature decrease further downstream, therefore, is caused solely by diffusion in the fluid. This peculiar situation arises because the back surface is insulated in this case. Close examination of Figures 13 and 14 will reveal that the thermal wake decays more rapidly for the turbulent flow cases, i.e., for velocities less than about 5 m/s (Re = 389200) for the smooth plate cases. It may also be seen that the onset of turbulent convection confines the region where substrate conduction is significant to no more than one length scale upstream from the source. Despite the conjugate nature of the problem, it is nevertheless tempting to define a local convective heat transfer coefficient as

$$ h = \frac{\dot{q}_{conv}}{T_s - T_o} = \frac{\dot{q}_{total} - \dot{q}_{cond} - \dot{q}_{rad}}{T_s - T_o} \tag{6} $$

and a local Nusselt number as

$$ Nu = \frac{hL_s}{k_f} = \frac{(\dot{q}_{total} - \dot{q}_{cond} - \dot{q}_{rad}) L_s}{(T_s - T_o) k_f} \tag{7} $$

Figure 16 shows that typically *Nu* is well behaved, decreases monotonically in the streamwise direction, and decreases monotonically with flow velocity, all consistent with expectations. Figure 17 demonstrates the dependence of the source average Nusselt number (averaged on the source area) on Reynolds number for a fixed source position of $x_s = 0.838$ m and for both smooth and grooved development surfaces. The results are surprisingly well behaved and readily interpreted. Regardless of the nature of the development surface, the data collect reasonably well on two paths: the upper path corresponds to the laminar regime, and the lower path corresponds to the turbulent regime. As seen previously in Figure 12, the transition from the laminar to turbulent behavior is dependent on the nature of the disturbances introduced by the surfaces. The best fit correlations through the data of Figure 17 are shown for the sole purpose of assessing the dependence on velocity.

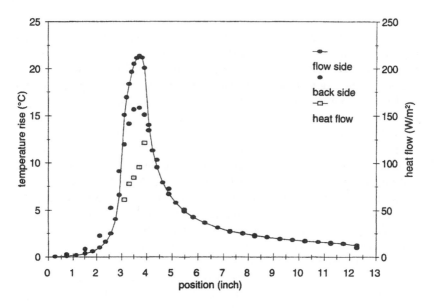

FIGURE 15. Surface temperature elevation above ambient on flow side and back side with smooth development plate, $x_s = 0.838$ m, $U = 5$ m/s, $Q_{total} = 0.5$ W/in.^{2}, [Ortega et al., 1994].

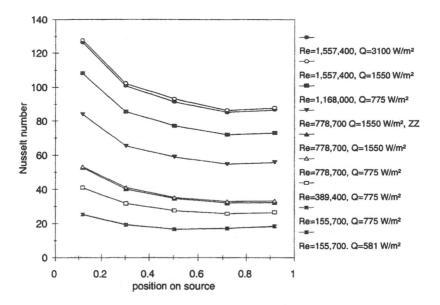

FIGURE 16. Local Nusselt number on source based on T_o and L_s, smooth development plate, $x_s = 0.838$ m, [Ortega et al., 1994].

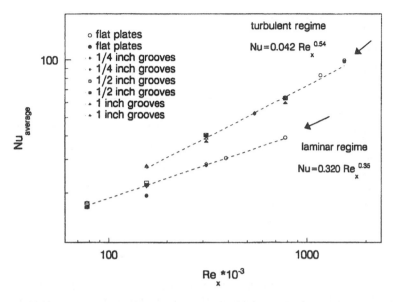

FIGURE 17. Average Nusselt number on source, $\bar{h}L_s/k_f$, dependence on Re_x, $x = x_s = 0.838$ m, [Ortega et al., 1994].

CONJUGATE MODELS
Laminar Channel Flow—Finite Difference/Control Volume Method

Numerous approaches for modeling the conjugate problem described by Figure 7 are possible, and many have been reported in the literature. Seminal work on the conjugate analysis of forced convection heat transfer from small isothermal heat sources embedded in large substrates for hydrodynamically fully developed laminar channel flow was performed by Ramadhyani et al. [1985] followed by the study of Incropera et al. [1986] on flush mounted isothermal heat sources embedded in one wall of the horizontal channel with hydrodynamically fully developed laminar or turbulent flow. Their efforts were concentrated on liquid cooling but their numerical approach was a general control-volume-based method in both the fluid and the solid. They conducted experimental studies and compared their results with a numerical investigation using a finite volume CFD approach in the fluid domain and a finite volume approach in the solid domains. Krause et al. [1989] repeated their study using a finite element method and investigated the effects of buoyancy. Sugavanam et al. [1995] performed a similar study for air and extracted many pertinent physical issues using the method. The approach of Sugavanam et al. is discussed next in some detail as it is representative of a full numerical formulation of the problem.

The domain of interest is shown in Figure 18 in which the strip source of heat is deployed on a conductive wall with laminar flow of air entering with

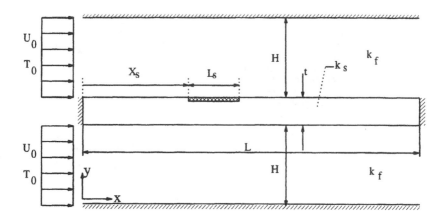

FIGURE 18.   The general two-dimensional conjugate problem [Sugavanam et al., 1995].

uniform velocity at the inlet to the channels formed by the substrate plate and the insulated walls above and below it. The exit conditions are that the stream-wise velocity and temperature gradients in the flow direction vanish at the exit plane. Unlike the experiment previously discussed in which the back side of the substrate surface was insulated, in this numerical experiment both the insulated backside and the convectively cooled backside cases were considered. For brevity, only results for the insulated backside conditions are discussed. The mathematical formulation for the case of the insulated backside is presented in brief. Steady, two-dimensional, Newtonian, incompressible, and laminar flow is assumed, and the properties of the material of the substrate and the fluid are considered to be uniform, isotropic, and constant. The primary variables are normalized as

$$X = \frac{x}{D_h} \quad Y = \frac{y}{D_h} \quad U = \frac{u}{U_0} \quad V = \frac{v}{U_0} \quad \theta = \frac{T - T_0}{\dot{q} D_h / k_f} \quad P = \frac{p}{\rho U_0^2} \tag{8}$$

$$Re = \frac{U_o D_h}{\nu} \quad Pr = \frac{\nu}{\alpha} \quad Pe = Re\,Pr \tag{9}$$

where $D_h = 2H$ is the channel hydraulic diameter and $\dot{q}$ is total power dissipation of the source per unit area.

With the above commonly applied assumptions, the continuity, x-momentum, y-momentum, and energy equations for the fluid side are given as

$$\frac{\partial U}{\partial X} + \frac{\partial V}{\partial Y} = 0 \tag{10}$$

$$U\frac{\partial U}{\partial X} + V\frac{\partial U}{\partial Y} = -\frac{\partial P}{\partial X} + \frac{1}{Re}\left[\frac{\partial^2 U}{\partial X^2} + \frac{\partial^2 U}{\partial Y^2}\right] \tag{11}$$

$$U\frac{\partial V}{\partial X} + V\frac{\partial V}{\partial Y} = -\frac{\partial P}{\partial Y} + \frac{1}{Re}\left[\frac{\partial^2 V}{\partial X^2} + \frac{\partial^2 V}{\partial Y^2}\right] \tag{12}$$

$$U\frac{\partial \theta}{\partial X} + V\frac{\partial \theta}{\partial Y} = \frac{1}{Pe}\left[\frac{\partial^2 \theta}{\partial X^2} + \frac{\partial^2 \theta}{\partial Y^2}\right] \tag{13}$$

and the corresponding boundary conditions at the inlet and exit are

$$U = 1, \quad V = 0, \quad \theta = 0 \quad \text{for} \quad X = 0, \quad \text{and} \quad \frac{t}{D_h} \leq Y \leq \frac{(H+t)}{D_h} \tag{14}$$

$$\frac{\partial U}{\partial X} = 0, \quad V = 0, \quad \frac{\partial \theta}{\partial X} = 0 \quad \text{for} \quad X = \frac{L}{D_h} \quad \text{and} \quad \frac{t}{D_h} < Y < \frac{H+t}{D_h} \tag{15}$$

Zero velocity, no-slip conditions are enforced at the interface and the upper wall, and the upper wall is assumed to be adiabatic, leading to

$$U = 0, \quad V = 0 \quad \text{for} \quad 0 \leq X \leq L/D_h \quad \text{and} \quad Y = \frac{t}{D_h}, \quad Y = \frac{H+t}{D_h} \tag{16}$$

$$\frac{\partial \theta}{\partial Y} = 0 \quad \text{for} \quad 0 \leq X \leq \frac{L}{D_h}, \quad \text{and} \quad Y = \frac{H+t}{D_h} \tag{17}$$

The Laplace equation describes the steady conduction in the solid side

$$\frac{\partial^2 \theta}{\partial X^2} + \frac{\partial^2 \theta}{\partial Y^2} = 0 \tag{18}$$

and the ends of the substrate are taken to be adiabatic, implying

$$\frac{\partial \theta}{\partial X} = 0 \quad \text{for} \quad X = 0, \quad X = \frac{L}{D_h}, \quad \text{and} \quad 0 \leq Y \leq \frac{t}{D_h} \tag{19}$$

$$\frac{\partial \theta}{\partial Y} = 0 \quad \text{for} \quad 0 \leq X \leq \frac{L}{D_h}, \quad \text{and} \quad Y = 0 \tag{20}$$

Finally and most critically, it is necessary to match heat flux and temperature at the interface between the fluid and solid domains

$$\frac{k_s}{k_f}\left[\frac{\partial \theta}{\partial Y}\right]_s = \left[\frac{\partial \theta}{\partial Y}\right]_f \quad \text{for} \quad 0 \leq X \leq X_s \quad \text{and} \quad X_s + \frac{L_s}{D_h} \leq X \leq \frac{L}{D_h} \tag{21}$$

$$\frac{k_s}{k_f}\left[\frac{\partial \theta}{\partial Y}\right]_s - \left[\frac{\partial \theta}{\partial Y}\right]_f = 1 \quad \text{for} \quad X_s \leq X \leq X_s + \frac{L_s}{D_h} \tag{22}$$

$$\theta_f = \theta_s, \quad \text{for} \quad 0 \leq X \leq \frac{L}{D_h} \quad \text{and} \quad Y = \frac{t}{D_h} \tag{23}$$

A numerical solution was found using the well documented simpler scheme described by Patankar [1980], and the details are available in Sugavanam [1994]. Unless otherwise specified, the results presented next are for the following nominal conditions: Reynolds number based on hydraulic diameter, Re_{Dh} = 1,260, corresponding to an average inlet velocity of 1.0 m/s, channel height H = 1 cm, heat source streamwise length L_s = 1 cm, and substrate thickness t = 0.5 cm. Variations of temperature, heat flux into the fluid, and Nusselt number along the interface, for a fully developed flow maintained at a constant Reynolds number and k_s/k_f varying between 0.1 and 100, are shown in Figures 19, 20, and 21. The source is located at a position $2.5 \leq X \leq 3.0$. The normalized temperature along the interface, Figure 19, remains undisturbed until the upstream conduction becomes nonnegligible. It then rises until it reaches a maximum on the heat source and decreases thereafter. In the wake region downstream of the source, the nondimensional temperatures for all conductivity ratios merge. It is especially instructive to learn that the downwind thermal wake achieves the adiabatic board behavior within only one length scale downstream, for $k_s/k_f \leq 10$, and within two length scales for greater conductivities. This is partly explained by the variation of the heat flux into the fluid side shown in Figure 20. For increasing k_s/k_f the spreading of the heat flux due to the board conduction is readily apparent, but the conduction is preferential to the upstream surface since the near wall fluid is cooler in this region. In the far downstream there is no heat flux into the fluid from the board, and hence the upstream distribution of the heat flux from the board into the fluid due to board conduction has little effect

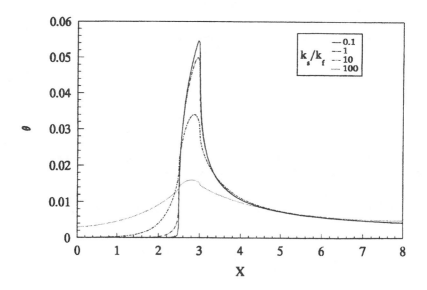

FIGURE 19. Interface temperature varying with position for fully developed flow at Re = 1260 for insulated board backside [Sugavanam, et al., 1995].

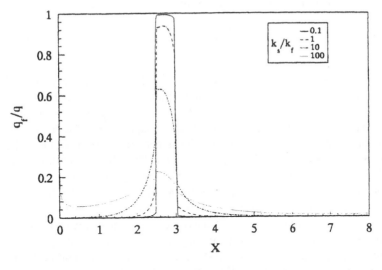

FIGURE 20. Heat flux into the fluid varying with position for fully developed flow at Re = 1260 for insulated board backside [Sugavanam et al., 1995].

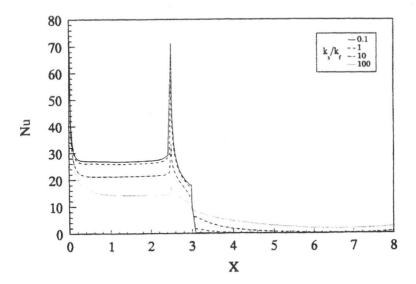

FIGURE 21. Nusselt number on the board-fluid interface as a function of position for fully developed flow at Re = 1260.0, insulated board backside [Sugavanam et al., 1995].

on the wall temperature. The wall approaches the true adiabatic wall temperature, for all k_s/k_f ratios. A local Nusselt number may be arbitrarily defined as

$$Nu = -\left[\frac{1}{\theta_f}\frac{\partial\theta_f}{\partial Y}\right]_i = \frac{\dot{q}_f D_h}{(T_s - T_0)k_f} \tag{24}$$

where $\dot{q}_f$ is the local convective heat flux into the fluid. The length scale used is arbitrarily chosen as the channel hydraulic diameter. As shown in Figure 21, the local Nusselt number is highly dependent on the conjugate board conduction. In the downstream wake region the heat flux approaches zero but the interface temperature elevation over the inlet temperature does not due to the convective wake effect, thus Nu approaches zero. The Nusselt number in the upstream region is a ratio of two small numbers, the heat flux from the board into the fluid and the temperature rise of the interface over the inlet. In the upstream region the ratio is nearly constant, but its value is dependent on the conductivity ratio. The difference in the behavior of Nu in the upstream region of the source compared to the downstream region is an excellent example of the difficulty in predicting Nu in heat transfer problems that have significant conjugate coupling and highly nonuniform surface thermal boundary conditions. Nu is lowest for the most conductive case as a result of heating of the near wall fluid upstream from the source through upwind conduction through the board. Increasing board conductivity results in a monotonic decrease of source averaged Nusselt number from its adiabatic board value as illustrated in Figure 22 for fully developed flow. This result is somewhat misleading in that the degra-

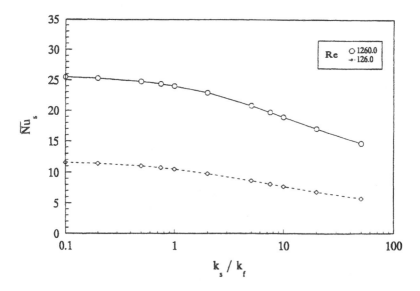

FIGURE 22. Source averaged Nusselt number dependence on conductivity ratio for fully developed flow, insulated board backside [Sugavanam et al., 1995].

dation of Nusselt number, when defined in terms of a fixed ambient temperature reference, appears to degrade the heat transfer from the heat source. In fact, increasing board conduction also increases the effective board area available for heat transfer and thereby decreases the heat source temperatures, a fact that is best observed in Figure 19. That fact notwithstanding, the source averaged Nusselt number may also be used to show the effects of the developing hydrodynamics, as illustrated by Figure 23. At a fixed board conductivity, the Nusselt number on the source decreases from its highest value at the inlet of the channel, where the boundary layer is thinnest, to its constant value under fully developed flow conditions. The asymptotic fully developed limits are indicated in Figure 23. Correlations were developed for the asymptotic fully developed flow Nusselt numbers as a function of Peclet number and board to fluid conductivity ratio. If $\overline{Nu}_{ad}$ is the Nusselt number for the adiabatic board case ($k_s/k_f = 0$), the correlational fit may be forced to yield the adiabatic board solution when k_s/k_f goes to zero. After a series of trials the following correlation emerged:

$$\frac{\overline{Nu}}{\overline{Nu}_{ad}} = \left\{ \frac{1.186}{Pe^{0.013}} \right\}^{-\left(\frac{k_s}{k_f}\right)^{0.586}} \tag{25}$$

where $Pe = RePr = U_oD_h/\alpha$. Equation 25 compared with the actual solution to within 1.2% for $0 \le k_s/k_f \le 10$. Alternatively, a simpler form valid for $0.1 \le k_s/k_f \le 10.0$ can be stated as

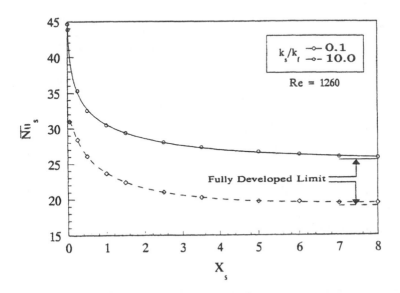

FIGURE 23. Source averaged Nusselt number dependence on position of source, laminar developing flow, Re = 1260, insulated board backside [Sugavanam et al., 1995].

$$\overline{Nu} = 1.833 \ Pe^{0.37} \left(\frac{k_s}{k_f} \right)^{-0.0754} \tag{26}$$

and this form collapsed the computed data to within 10%. Both Equations 25 and 26 are for the single substrate thickness, $t/L_s = 0.5$.

An important virtue of the single source solutions is that they may be used to find the solution for a situation of multiple sources on a conducting substrate, interacting both convectively and conductively through the board by linearly superposing the single source kernel solutions. The energy equations both on the solid side and the fluid side are linear with respect to temperature, thus allowing the superposition. This fact is verified numerically for two discrete, identical, flush-mounted, constant-powered heat sources embedded in a conductive substrate for the case where the backside of the board is insulated. Figure 24 demonstrates the comparison between the superposed interface temperature solution and the actual temperature solution for two discrete, flush-mounted sources under fully developed flow conditions and k_s/k_f of 10. The length of the channel is 17 cm with the first source located at 5 cm from the inlet and the second source spaced 5 cm from the first source. Perfect agreement is observed as expected. It can be concluded that the solution for a single isolated source of heat on a conducting substrate may be used as a kernel solution thus reducing the computational time required for a thermal design tool.

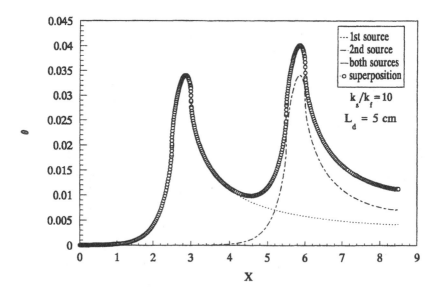

FIGURE 24. Interface temperature varying with position for fully developed flow at Re = 1260, for two sources with distance between sources, $L_a = 5$ cm, $k_s/k_j = 10$, insulated board backside [Sugavanam et al., 1995].

Laminar Channel Flow—Boundary Element Method

The finite difference/volume approach discussed in the previous section is by far the most commonly used method in commercially available CFD codes that allow solution of conjugate problems. The accuracy of the approach may suffer in regions of high gradients because of improper grid resolution. In particular for the problems of the type described in this chapter, proper computation of the heat flux at the fluid–solid boundary, especially in regions of rapid changes such as in the vicinity of heat sources, is difficult using a finite volume approach. A powerful alternative approach, using the boundary element method (BEM) in the solid domain, is computationally superior for the two-dimensional problem because it allows the computation of both temperature and heat flux as primary variables [Kabir *et al.*, 1995] and it can be readily implemented with a variety of flow solvers. Because the approach is not well-known in the electronics thermal management community, a brief description of its implementation follows, and results for the two-dimensional problem of Figure 25 for fully developed channel flow are discussed and compared to the results of the finite-volume approach.

Solid Side Model

In the boundary element method, only the boundaries of the domain of interest are discretized. This is accomplished by the transformation of the governing differential equation into boundary integral equations. As stated earlier, the accuracy of the BEM solutions is greater than those obtained using finite element or equivalent methods because the temperature (potential) and heat flux density (gradient) are directly calculated, thus retaining the same accuracy

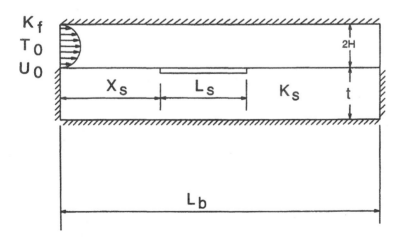

FIGURE 25. Problem domain for BEM analysis of two-dimensional conjugate problem [Kabir et al., 1995].

for both. In contrast, finite element results are accurate for the primary variable (potential) of the field problems. The gradients are obtained by numerically differentiating the potentials. In doing so, the heat flux densities are much less accurate and are usually discontinuous between elements [Lee and Palisoc, 1990]. Referring to the general domain of Figure 26, the steady state heat conduction equation for an isotropic solid domain Ω bounded by the surface Γ is

$$\nabla^2 T = 0 \quad \in \quad \Omega \tag{27}$$

subject to the boundary conditions

$$T = T_0 \quad on \quad \Gamma_D \tag{28a}$$

$$\frac{\partial T}{\partial n} = q = q_{n0} \quad on \quad \Gamma_N \tag{28b}$$

$$aT + bq = c \quad on \quad \Gamma_M \tag{28c}$$

where T is the temperature and q is the normal heat flux density. Subscript O denotes the prescribed values and subscripts D, N, and M denote boundaries with Dirichlet, Neumann, and mixed boundary condition types respectively;

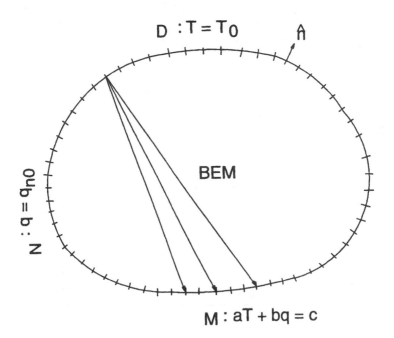

FIGURE 26. BEM elements along the boundary of a typical domain indicating three types of possible boundary conditions [Kabir et al., 1995].

n is the outward normal to the boundary. The constants a, b, and c are known quantities (a and b both are nonzero).

Multiplying Equation 27 by a weighting function w and integrating by parts results in

$$\int_\Omega T\nabla^2 w\, d\Omega + \int_\Gamma (wq - Tw_n)\, d\Gamma = 0 \tag{29}$$

in which subscript n denotes the normal derivative. Now w is chosen as the free space Green's function T^* satisfying

$$\nabla^2 T^*(\bar{x}) = \delta(\bar{x}) \tag{30}$$

which in two dimensions has the solution

$$T^* = \frac{1}{2\pi} \ln\left(\frac{1}{R}\right) \tag{31}$$

R being the distance between the field point and the source point. Merging of Equation 30 into Equation 29 and taking the limit as $\bar{x}$ tends to any point p on the boundary results in

$$c(p)T(p) = \int_\Gamma (qT^* - Tq^*)\, d\Gamma \tag{32}$$

The coefficient c is a function of the internal angle the boundary makes at point p (local geometry). If the boundary is locally smooth c has the value of $1/2$. Otherwise, it may be calculated directly or indirectly [Brebbia et al., 1984]. Now Equation 32 is discretized by dividing the boundary into N number of elements as shown in Figure 26, while assuming that the value of T and q at a point within an element is related to its values at the element nodal points by some interpolation function. If linear interpolation functions are chosen for both temperature and normal flux densities, Equation 32 becomes

$$c_i T_i = \sum_{j=1}^N \int_{\Gamma_j} (\psi_1 q_1 + \psi_2 q_2) T^*\, d\Gamma - \sum_{j=1}^N \int_{\Gamma_j} (\psi_1 T_1 + \psi_2 T_2) q^*\, d\Gamma \tag{33}$$

where

$$\psi_1 = \frac{1}{2}(1 - \eta); \quad \psi_2 = \frac{1}{2}(1 + \eta)$$

and η is a dimensionless coordinate equal to $2y/l$, l being the element length and y a local coordinate. T_1, T_2, q_1, q_2 are the nodal quantities of the jth element along the boundary. Now defining

$$g_{ij}^k = \int_{\Gamma_j} \psi_k T^*\, d\Gamma \quad ; \quad h_{ij}^k = \int_{\Gamma_j} \psi_k q^*\, d\Gamma - c_i \delta_{ij}$$

and adding the contributions from the two neighboring elements, (j-1) and (j), into one term, Equation 33 can be rewritten in the form

$$\sum_{j=1}^{N} H_{ij}T_j = \sum_{j=1}^{N} G_{ij}q_j \tag{34}$$

where N is the total number of boundary nodes and each nodal coefficient G_{ij} is equal to the term g_{ij}^2 of element $(j-1)$ plus the term g_{ij}^1 of element (j), for a counterclockwise numbering system. The same applies for H_{ij}. For a well-posed problem, either T or q or some relation between the two is known for each node. Rearranging Equation 34, a set of N linear equations with N unknowns are found as described by the following equation, which can be readily solved for nodal unknown quantities.

$$\sum_{j=1}^{N} A_{ij}Y_j^{(u)} = \sum_{j=1}^{N} B_{ij}Y_j^{(k)} \tag{35}$$

The superscripts u and k mean unknown and known quantities, respectively.

Fluid Side Model

As in any conjugate problem, the interface temperature distribution or heat flux uniquely determines the temperature field in the fluid; however, neither can be specified *a priori*. In principle, any model of the fluid side energy equation that can be posed in terms of the interfacial temperature or heat flux can be coupled to the solid side formulation. To illustrate, an analytic approach proposed in this context by Culham et al. [1991a] is used, both to contrast it with the full CFD approach of Sugavanam et al. [1995] and to illustrate the power of analytical approaches for the practical problems in air cooling of electronics.

It is not necessary to solve the fluid flow problem explicitly to formulate the conjugate heat transfer if the *response* of the flow in question due to a step change in either wall heat flux or wall temperature is available in analytical form. The reason is that any arbitrary distribution of surface temperature or heat flux may be represented by the superposition of a series of step changes, (see, for example, the discussion of Eckert and Drake [1972], pp. 321–327). In the conjugate problem, when the solid side is discretized, we may consider that the objective is to develop an expression implicitly relating the fluid temperature at the wall to the wall heat flux in each discrete interval, where neither is known *a priori*. For the present two-dimensional problem it is straightforward to illustrate. For a surface in which the boundary condition specified is some spatially varying temperature, $T_w(x)$, the local heat flux $\dot{q}_f(x)$ can be evaluated using the Duhamel superposition technique, first proposed by Rubesin [1951]

$$\dot{q}_f(x) = \int_{\xi=0}^{x} h(\xi,x) \frac{dT_w(\xi)}{d\xi} d\xi \tag{36}$$

where $h(\xi,x)$ is the unit thermal conductance, which may be found from the solution of the relevant energy equation in the presence of a step function in temperature. In this case, x is the streamwise spatial position, $0 \leq x \leq L_b$, and ξ is a local dummy variable, $0 \leq \xi \leq x$ accounting for the evolution of the temperature in the flow up to the local position x. Solutions for $h(\xi, x)$ are documented for many types of flows. For example, integral formulations of the laminar boundary layer solutions were collected by Rubesin [1951] and Tribus and Klein [1952], and these are discussed in a subsequent section. In the present case, for fully developed laminar channel flow, the solutions of McCuen [1962] for $h(\xi,x)$ are perfectly suitable. The unit thermal conductance is

$$h(\xi, x) = \frac{k_f}{H} \sum_{n=0}^{\infty} C_n Y_n' \exp\left\{-\frac{32}{3D_h RePr} \lambda_n^2 (x - \xi)\right\} \tag{37}$$

where C_n are eigenconstants, λ_n are the eigenvalues, and Y_n' are the derivatives of eigenfunctions evaluated at the wall where the step in temperature is applied. Equation 36 is discretized along the solid-fluid interface in the streamwise direction by dividing the interface along the streamwise direction into $(k-1)$ elements, as shown in Figure 27

$$\dot{q}_f(x_j) = \sum_{i=1}^{j-1} \left(\frac{T_w(x_{i+1}) - T_w(x_i)}{x_{i+1} - x_i}\right) \int_{x_i}^{x_{i+1}} h(\xi, x) d\xi, \quad j = 1, 2, \ldots k \tag{38}$$

All the coefficients of T_i are evaluated numerically using standard quadrature except for the case when $x_i = x_j$, for which a closed form solution exists. In matrix form Equation 38 can be rewritten as

$$\dot{q}_f(x_j) = \sum_{i=2}^{j} \alpha_{ji} T_w(x_i) - c_j \quad j = 2, 3, \ k \tag{39}$$

The c_j's are related to $\dot{q}_f(x_j)$, and are due to $T_w(x_i)$.

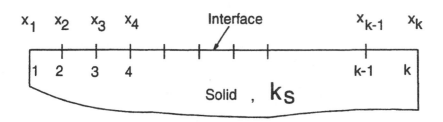

FIGURE 27. Discrete along the fluid-solid interface when applying BEM scheme [Kabir et al., 1995].

Solid-Fluid Coupling

The coupling between the solid and the fluid region is accomplished by ensuring the energy balance and matching the temperatures at the interface. That is, at any node j, along the interface

$$T_w(x_j) = T_j \tag{40}$$

$$\dot{q}_f(x_j) + k_s q_j = Q_j \tag{41}$$

For nonsource nodes, Q_js are zero. The BEM model, Equation 35, can be interpreted as a relation between the unknown $Y_j^{(u)}$ and the known (specified) conditions. The change in the heat flux density at the jth node, due to a change in temperature at the lth node, can be obtained by taking the derivative of Equation 35 with respect to T_i

$$\sum_{j=1}^{N} B_{ij} \frac{\partial q_j}{\partial T_l} = A_{il}. \tag{42}$$

From Equation 42 it is evident that all the derivatives are independent of temperature and flux densities, hence, an exact linear relation can be derived to calculate the correct flux densities

$$q_j - q_j^{(0)} = \sum_{l=2}^{k} (T_l - T_l^{(0)}) \frac{\partial q_j}{\partial T_l} \quad j = 2,3, \ k \tag{43}$$

Here $q_j^{(0)}$ is the solution of Equation 35 based on an arbitrary prescribed $T_l^{(0)}$. It should be pointed out that, depending on the desired heat flux condition, the correct temperature T_l can be determined exactly from Equation 43 without any iteration. Multiplying Equation 43 by k_s and adding the resulting equation to Equation 39, Equation 41 can be used to arrive at the following:

$$\sum_{j=2}^{k} \tilde{A}_{ij} T_j = D_i \tag{44}$$

where

$$\tilde{A}_{ij} = \alpha_{ij} + k_s \frac{\partial q_i}{\partial T_j}, \quad j \le i$$

$$\tilde{A}_{ij} = k_s \frac{\partial q_i}{\partial T_j}, \quad j > i$$

$$D_i = \sum_{l=2}^{k} k_s \frac{\partial q_i}{\partial T_l} T_l^{(0)} + Q_i + c_i - k_s q_i^{(0)}$$

Equation 44 now can be solved readily to obtain the interface temperature distribution and either Equation 35 or Equation 39 to calculate the heat flux distribution.

For comparison, the geometry (Figure 25) was chosen to match that of Sugavanam et al. [1995]; in the notation of Figure 25, $2H/L = 1$, $2H/t = 2$, $L_s = 1$ cm, $t = 0.5$ cm, and $L_b = 16$ cm, and the nondimensional temperature was defined as

$$\theta = \frac{(T - T_0)}{Q4H} k_f.$$

(45)

Figure 28 compares the BEM computed results with those of Sugavanam for a nominal velocity of 1 m/s, $Re = 1260$. There is nearly perfect agreement between the two techniques. The computed convective heat flux, given in Figure 29, also agrees reasonably well with the previous results shown in Figure 20. The BEM method is probably more accurate in the vicinity of the source leading edge since it shows continuous, smooth behavior, whereas the finite volume method, Figure 20, shows discontinuous behavior, especially at low conductivity ratios. When comparing local Nu at the interface, Figure 30, more obvious differences can be seen. Whereas the finite volume approach predicts the Nu to approach a constant value upstream of the source, different for each conductivity ratio, the BEM approach predicts this behavior only at relatively high conductivity ratio; otherwise, the Nu starts at zero and rises to its peak value at the leading edge of the source, but with an unusual intermediate inflection point. We believe that the BEM approach captures the true behavior of Nu because

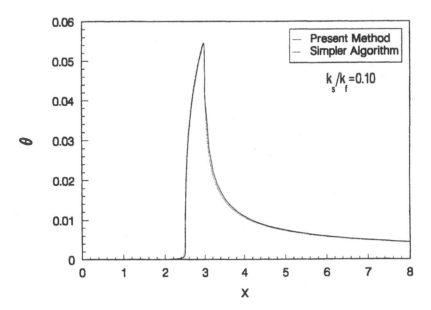

FIGURE 28. Interface for fully developed channel flow; comparison of BEM result to result of Sugavanam et al., [1995] using finite volume method [Kabir et al., 1995].

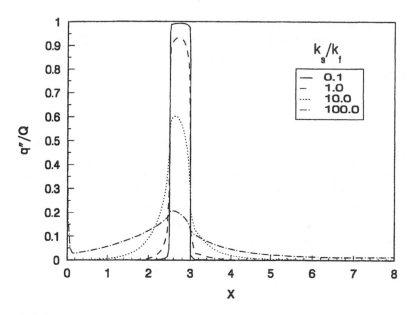

FIGURE 29. Convective heat flux into fluid q'' computed with BEM approach, normalized by total heat source power per unit area, Q, fully developed channel flow, Re = 1260 [Kabir et al., 1995].

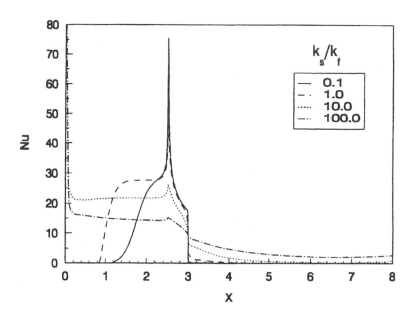

FIGURE 30. Nusselt number distribution along interface computed with BEM approach, fully developed channel flow, Re = 1260 [Kabir et al., 1995].

both wall heat flux and temperature are computed directly. The finite volume method requires that the wall heat flux be determined by numerically evaluating the normal gradient, and therefore its accuracy degrades when both heat flux and temperature rise approach zero. On the other hand, the BEM method is not as straightforward to extend to three dimensions as either the finite-difference/volume or the finite element method, and this is a major drawback.

Laminar Boundary Layer Flow

Given the previous discussion on the flow regimes in typical PCB applications, an equally important flow to consider is that of laminar boundary layer flow. The domain of interest is represented in Figure 31. Again, various approaches for the conjugate problem can be taken and several are briefly illustrated here.

Using a finite difference, control volume approach similar to that of Sugavanam et al. [1995], Gorski and Plumb [1990, 1992] analyzed the conjugate behavior of the two-dimensional strip heat source problem [1990] and the three-dimensional rectangular source problem [1992] with the flow described by the analytical solution of Blasius for the laminar boundary layer. They considered only the limiting case of infinitely thick substrate. In the domain illustrated by Figure 25, they solved the energy equation for the flow side, Equation

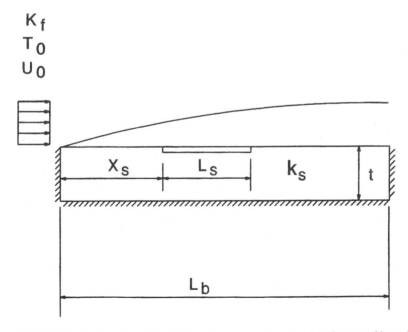

FIGURE 31. Problem domain for BEM analysis of two-dimensional conjugate problem with laminar boundary layer flow [Kabir et al., 1995].

13, using a finite-difference method with the velocity field given by the Blasius solution at each streamwise position. In the solid side, the energy equation, Equation 18, was discretized also with a finite-difference method, and the entire system of discretized equations in both domains was simultaneously solved using a line-by-line algorithm with temperatures along each line evaluated using matrix inversion. The boundary and matching conditions were the same as those previously formulated, and the discontinuity in the thermal conductivity at the fluid/solid interface was handled using the harmonic mean technique suggested by Patankar [1980]. Their results were very similar to those of Sugavanam et al. [1995] but cannot strictly be compared because the developing channel flow adds freestream acceleration to the boundary layer. Gorski and Plumb conducted a systematic variation of parameters, varying Peclet number, $Pe = (U_o x_s) \alpha$ between 1×10^3 and 1×10^5, conductivity ratio, k_s/k_f between 0.1 and 10.0, and streamwise position to source-length, x_s/L_s between 5 and 100. Their source averaged Nusselt number, $\overline{Nu} = \overline{h} L_s/k_f$ was correlated to within 5% by the expression

$$\overline{Nu} = 0.486 Pe^{0.53} \left(\frac{L_s}{x_s}\right)^{0.71} \left(\frac{k_s}{k_f}\right)^{-0.057} \tag{46}$$

The Nusselt number was defined in terms of the source length, L_s, and Peclet number was defined in terms of streamwise position of the source, x_s, in contrast to the correlations of Sugavanam, Equations 25, 26 in which the channel hydraulic diameter was used for both Nu and Pe. As in the previous case for fully developed channel flow, increase in substrate conductivity decreased the source Nusselt number. In addition, the boundary layer growth decreased the average Nusselt number as shown by its inverse dependence on source position, varying approximately as $x^{-0.18}$. This is consistent with the channel flow problem, Figure 23, in which Nu decreases in the channel entry region to its fully developed value. In some instances it might be more useful to average the heat transfer coefficient or Nusselt number over the entire substrate area rather than just over the source. For example, if one were to perform a conjugate analysis of the heat transfer in the plastic monochip package shown in Figure 5, it might be more meaningful to determine the average heat transfer coefficient over the entire cooled top surface area. In the present cases, an average heat transfer coefficient is extracted for the source area rather than for the entire board.

Alternatively, the BEM approach, Equations 27 to 44, can be used with the laminar boundary layer flow provided a suitable expression is used for the unit surface conductance, Equation 36. The exact solution for boundary layer flow is cited in Eckert and Drake [1972] as

$$h(\xi, x) = \frac{0.332 k_f}{x} \ Pr^{\frac{1}{3}} Re_x^{\frac{1}{2}} \ \left\{ 1 - \left\{ \frac{\xi}{x} \right\}^{\frac{3}{4}} \right\}^{-\frac{1}{3}} \tag{47}$$

Culham et al. [1991a] showed that a general form of Equation 47 can be written as

$$h(\xi, x) = C \frac{k_f}{x} \; Pr^\gamma Re_x^\beta \; \left\{ 1 - \left(\frac{\xi}{x} \right)^a \right\}^{-b} \tag{48}$$

where the parameters, C, γ, β, a, and b have been derived for various types of flows, including the present laminar boundary layer flow. Table 3, abstracted from their paper, summarizes some of the available solutions.

Using Equation 47, results were obtained by Kabir et al. [1995] using the BEM approach in the solid, for k_s/k_f between 0.1 and 100 and for free stream velocity of 1.0 m/sec, equivalent to $Re_{L_b} = 10,000$. Results were very similar to those for fully developed channel flow. Figure 32 shows that local Nusselt number behaves in much the same way, with the intermediate inflection point in the upstream region occurring at low board conductivity. By comparison with Figure 23, for the fully developed channel flow, and Figure 33, for the boundary layer flow, one may note that for both conductivity ratios, Nu for the channel problem is significantly higher, presumably because the growth rate of the unconstrained boundary layer is greater than the growth rate of the channel flow. In addition, the source Nusselt number does reach a fully developed asymptote. For low to moderate conductivity ratios, the thermal boundary layer thickness over the source is less than half the channel height so that the thermal boundary conditions on the opposing wall have no influence on the source heat transfer.

TABLE 3
Constants To Be Used in Equation 48 for Boundary Layer Flow

Solution	C	γ	β	a	b	Reference
			laminar			
exact	0.332	1/3	1/2	3/4	1/3	Eckert [1950]
approx	0.304	1/3	1/2	3/4	1/3	Rubesin [1945]
approx	0.323	1/3	1/2	3/4	1/3	Kays [1993]
			turbulent			
approx	0.0288	1/3	4/5	39/40	7/39	Rubesin [1951]
approx	0.0289	1/9	4/5	9/10	1/9	Seban [1950]

Source: Culham et al., 1991a. With permission.

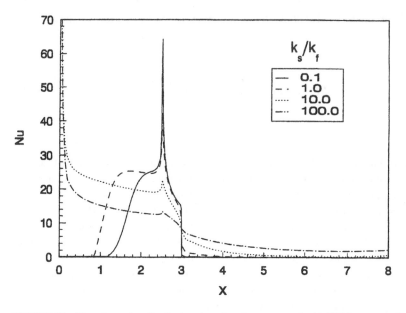

FIGURE 32. Nusselt number distribution along interface computed with BEM approach, laminar boundary layer flow, $Re_{Lb} = 10,000$, source position $x_s/L_b = 5/16$ [Kabir et al., 1995].

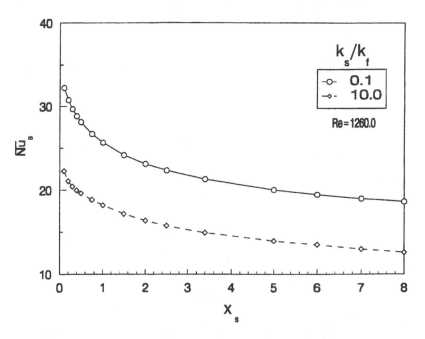

FIGURE 33. Average Nusselt number on source computed with BEM approach, dependence on streamwise position of source, laminar boundary layer flow, $Re_{Lb} = 10,000$ [Kabir et al., 1995].

THREE-DIMENSIONAL SITUATIONS: RECTANGULAR SOURCES OF HEAT

We next consider selected approaches and results for the problem of Figure 6(b), the cooling of a rectangular, flush-mounted source of heat on a conducting substrate or board. Figure 34 indicates the geometry with the pertinent geometric parameters. Although the system ignores the difficulties associated with more difficult flow, as for example in Figure 6(f), the fact that the flow is well-known facilitates the understanding of the conjugate phenomena. In addition, as demonstrated by Culham et al. [1991b], at low laminar Reynolds numbers, the flush-source model may give reasonable agreement with measured data on PCBs.

BOUNDARY LAYER FLOW
In the coordinate system given in Figure 34, the boundary layer equations for mass and momentum are commonly given as

$$\frac{\partial u}{\partial x} + \frac{\partial v}{\partial z} = 0 \tag{49}$$

$$u\frac{\partial u}{\partial x} + v\frac{\partial u}{\partial z} = v\frac{\partial^2 u}{\partial z^2} \tag{50}$$

and the energy equation, retaining molecular diffusion terms in all three directions is

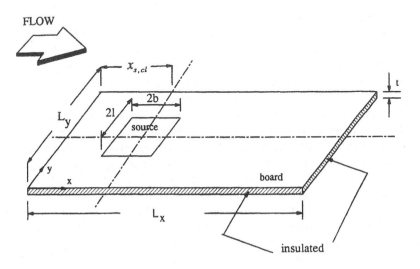

FIGURE 34. Geometry for boundary layer flow over a rectangular heat source on a conducting substrate or board.

$$u\frac{\partial T}{\partial x} + \upsilon\frac{\partial T}{\partial z} = \alpha_f\left(\frac{\partial^2 T}{\partial x^2} + \frac{\partial^2 T}{\partial y^2} + \frac{\partial^2 T}{\partial z^2}\right) \tag{51}$$

In general only the last term on the right, representing diffusion normal to the substrate surface, is retained for the low speed laminar boundary layer, but all three terms are retained here for initial discussion. The energy equation in the solid is in general

$$\frac{\partial^2 T}{\partial x^2} + \frac{\partial^2 T}{\partial y^2} + \frac{\partial^2 T}{\partial z^2} + \frac{S}{k_s} = 0 \tag{52}$$

where S is a volumetric source term that allows specification of heat dissipation at the location of the source.

Gorski [1992] and Gorski and Plumb [1992] extended their finite-difference, control-volume-based approach discussed in the previous section to the cooling of a rectangular source on an infinitely thick substrate in a laminar boundary layer flow. Rather than solving the laminar boundary layer equations 49 and 50, they adopted the classical solution of Blasius for the velocity components. The energy equations 51 and 52 were discretized using a finite-difference, control-volume approach with a staggered grid. The discontinuity in thermal conductivity at the fluid/solid interface was handled using the harmonic mean suggested by Patankar [1980]. The edges of the board were assumed insulated and the substrate thickness was chosen to be large enough such that infinitely thick substrate behavior was achieved. The substrate surface area was assumed to be large enough to allow the fluid temperature at the leading edge and at the lateral edges to be taken as the undisturbed temperature T_o. At the outflow boundary, the fluid temperature gradient in the streamwise direction was assumed to be zero. Results were obtained for $0.1 \leq k_s/k_f \leq 10$, $5 \leq (x_{s,cl})/2b \leq 150$, $0.2 \leq l/b \leq 5.0$, and $1 \times 10^3 \leq Pe \leq 1 \times 10^5$, where Peclet number for this case is $Pe = U_o x_{s,cl}/\alpha$. Representative results are shown in Figures 35 and 36. Figure 35 demonstrates the dependence of temperature on the spanwise centerline of the source on board conductivity at a fixed $Pe = 5 \times 10^3$. At the lowest conductivity, in which the substrate is nearly adiabatic, the temperature of the square source nearly matches that of the two-dimensional strip source, indicating that spanwise conduction either through the substrate or fluid is negligible compared to streamwise advection. As conductivity ratio increases, the square and strip source temperatures substantially diverge, even for $k_s/k_f = 1.0$, in which the solid conductivity would be that of air. This divergence would decrease at higher Pe in which streamwise advection would dominate the heat transfer. For example, Figure 36 shows that for a conductivity ratio of 1.0, but at $Pe = 7.5 \times 10^4$, the square source, $l/b = 1.0$, more nearly approaches the two-dimensional strip source behavior. For $l/b \geq 5.0$, the square source temperatures substantially match the two-dimensional solution. Based on the data generated using their model, the authors developed the following correlations for average source Nusselt number, $\overline{Nu} = \overline{h}2b/k_f$

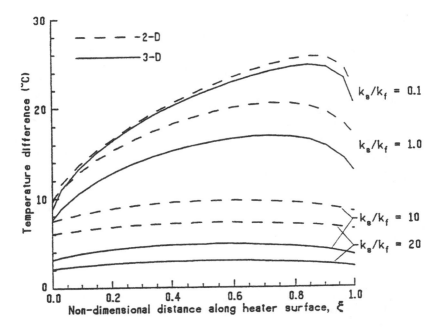

FIGURE 35. Effect of the board conductivity on the spanwise centerline temperature for a square source compared to strip source of same streamwise length in a laminar boundary layer, $Pe = U_o x_{s,cl}/\alpha = 5 \times 10^3$, $x_{s,cl}/2b = 25$, $l/b = 1.0$ [from Gorski, 1992].

$$\overline{Nu} = 0.60 Pe^{0.48} \left(\frac{2b}{x_{s,cl}}\right)^{0.63} \left(\frac{b}{A/P}\right)^{0.18} \quad ; \quad k_s/k_f = 1.0 \tag{53}$$

$$\overline{Nu} = 0.43 Pe^{0.52} \left(\frac{2b}{x_{s,cl}}\right)^{0.70} \left(\frac{b}{A/P}\right)^{0.07} \quad ; \quad k_s/k_f = 10.0 \tag{54}$$

where A/P is the source surface area to perimeter ratio. In order to correlate the source geometry, both the aspect ratio l/b and the source surface area to perimeter ratio normalized on source length $(A/P)/b$ were tried with the latter giving slightly better fit to the data. Since in practice the total heat dissipated Q is generally known, it is useful for practical calculations to correlate the total heat transferred to the fluid Q_f. The authors found the following best fits of their numerical data

$$\frac{Q_f}{Q} = 0.36 \left\{ Pe \left(\frac{2b}{x_{s,cl}}\right)^{1.5} \left(\frac{A/P}{b}\right)^{1.8} \right\}^{0.19} \quad ; \quad k_s/k_f = 1.0 \tag{55}$$

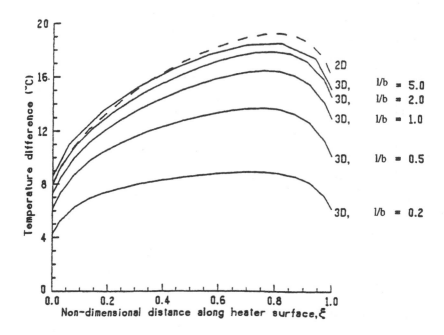

FIGURE 36.   Effect of aspect ratio l/b on the spanwise centerline temperature of a rectangular source in a laminar boundary layer, $Pe = U_o x_{s,cl} \alpha = 7.5 \times 10^4$, $x_{s,cl}/2b = 100$, $k_s/k_f = 1.0$ [from Gorski, 1992].

$$\frac{Q_f}{Q} = 0.063 \left\{ Pe \left(\frac{2b}{x_{s,cl}} \right)^{1.5} \left(\frac{A/P}{b} \right)^{1.8} \right\}^{0.36} \quad ; \quad k_s/k_f = 10.0 \qquad (56)$$

In contrast to the full numerical approach of Gorski and Plumb, Culham et al. [1991a, 1991b] developed an analytical/numerical model for the problem of Figure 34 for a board with finite thickness. The board was assumed to be thin enough so that normal (z-direction) temperature gradients could be neglected. This "thin board" model was justified for board Biot number $Bi = ht/k_s < 0.1$, which includes most practical applications of air-cooled PCBs. Neglecting normal temperature gradients allows integration of Equation 52 across the thickness of the board to yield the thin board result

$$\frac{\partial^2 T}{\partial x^2} + \frac{\partial^2 T}{\partial y^2} + \frac{1}{k_s t}(q_s - q_f - q_r) = 0 \qquad (57)$$

where q_s is local power dissipation per unit surface area, q_f is the convective heat flux into the fluid, and q_r is the radiative heat flux from the board surface. Both q_f and q_r were assumed to be the sum of the fluxes from the upper and lower surfaces. The lateral edges of the board were assumed to be insulated.

The solid-side solver was built on a finite-element based approach. Culham et al. [1991a] introduced the solution to the flow-side energy equation by adopting the unit surface conductance method discussed previously in the context of the BEM. If diffusion in the fluid is neglected in the streamwise (x) and spanwise (y) directions, Equation 51 becomes

$$u\frac{\partial T}{\partial x} + v\frac{\partial T}{\partial z} = \alpha_f \frac{\partial^2 T}{\partial z^2} \tag{58}$$

which is the common boundary layer form of the energy equation. Again, under conditions of a temperature-specified wall, the local heat flux is given by Equation 36, where the form of the unit thermal conductance h (ξ,x) is dependent on the type of flow. For boundary layer flow, Culham et al. [1991a] proposed the general form given by Equation 48. Substitution of Equation 48 into Equation 36 gives

$$q_f(x) = \int_{\xi=0}^{x} C\frac{k_f Pr^\gamma}{x^{(1-\beta-ab)}}\left(\frac{\rho U_o}{\mu}\right)^\beta [x^a - \xi^a]^{-b}\theta_w(\xi)\,d\xi \tag{59}$$

where the constants were previously tabulated in Table 3. It was further demonstrated that Equation 59 can be inverted so that the wall temperature elevation, $\theta_w(x) = T_w(x) - T_o$, can be expressed explicitly for conditions of a heat flux specified boundary as

$$\theta_w(x) = \frac{Pr^{-\gamma}Re_x^{-\beta}x}{\Gamma(b)\Gamma(1-b)(Ck_f)}\int_0^1 q_f[1-\chi^a]^{-(1-b)}\,d\chi \tag{60}$$

where Γ is a gamma function. This form can be simplified if q_f is constant over a discrete interval, which is the form adopted by Culham et al. in coupling to the board FEM equations.

It is worth emphasizing that these solutions neglected both streamwise and spanwise conduction in the fluid resulting therefore in a solution that is functionally dependent only on the streamwise position. In the absence of conjugate coupling to the board, this solution could not be used in the three-dimensional problem, because no mechanism would exist for spanwise diffusion of the thermal energy. However, when coupled to the solid-side solution that allows conduction in the board in both the streamwise and spanwise directions, the two-dimensional solution represented by Equations 59 or 60 may be used, with the implicit assumption that the magnitude of spanwise or lateral conduction in the board renders the spanwise diffusion in the fluid negligible. Obviously, this assumption can be true only for $k_s/k_f > 1.0$, but this includes all the cases of practical importance. Note however that this fluid side solution would not be useful in the case of a poorly conducting or adiabatic board.

The finite-element model in the board allows discretizing of the board into finite areas sufficiently small to allow the surface heat flux calculated at the mid-point to be representative of the heat flux over its area. The unit step function models for the fluid side and the FEM models on the solid side were coupled by iterating between each solution until the calculated wall heat flux distribution forced the wall temperature calculated using the fluid flow equations to match the wall temperature calculated with the solid side model. The iterative procedure was instituted in a general purpose code for the PCB problem [Culham et al., 1991a], and a detailed parametric study was performed [Culham et al., 1991b] on the effects of board emissivity, board thermal conductivity, and for the two source model illustrated in Figure 37, effects of upstream source heating and intersource spacing on the downstream source. Nomenclature is shown in Figure 37. The dimensionless temperature on the spanwise centerline for the two source test case studied is presented in Figure 38, where

$$\theta = \frac{T - T_o}{\bar{q}_j \, L/k_f} \tag{61}$$

and the average heat flux over the front and back surface of the board is defined as

$$\bar{q}_j = \frac{\Sigma Q_i}{2(L x W)} \tag{62}$$

As board conductivity gets very large, heat spreading renders the board nearly isothermal. The board conductivity influences the peak temperature on

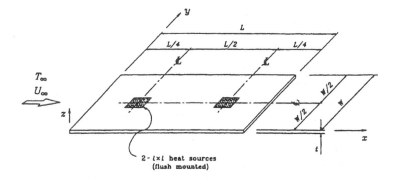

BASE CONDITIONS

$L \times W \times t = 0.2m \times 0.1m \times 0.0016m$ $U_\infty = 5m/s$

$k_s = 2W/m - K$ $T_\infty = 20C$

$l \times l = 0.02m \times 0.02m$ $Q_1 = Q_2 = 2W$

FIGURE 37.   Geometry and nomenclature for two source PCB [from Culham et al., 1991b].

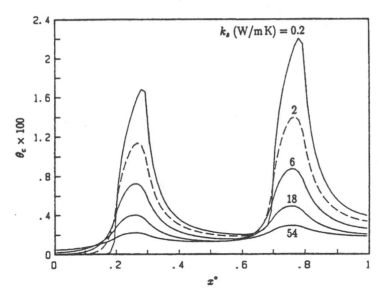

FIGURE 38. Spanwise centerline temperature dependence on board thermal conductivity, $Q_1 = Q_2 = 2.0$ W, [from Culham et al., 1991b].

both sources, and the convective thermal wake mechanism increases the temperature of the downstream source above the upstream source for all conductivities, the relative difference depending on the intersource spacing. In a separate study [Lemczyk et al., 1991], it was found that an accurate method for determining the effective thermal conductivity of a multilayer printed circuit board is by use of the harmonic mean of the thermal conductivity of individual layers calculated using appropriate weighting of the series and parallel resistances. Figure 39 shows the effect of the strength of the upstream source on the temperature of the downstream source. For the intersource spacing shown, about 90% of the temperature increase of the downwind source is due to its own heating, while 10% is due to the influence of the upwind source. The upwind source temperature increases linearly with its own power dissipation. Effect of intersource spacing is shown more generally in Figure 40, where the temperature increase of the second source ΔT_2 above its base value for the isolated case, $\Delta T_{2,o}$, is shown. The ratio ϕ is

$$\phi = \frac{\Delta T_2 - \Delta T_{2,o}}{\Delta T_{2,o}} \tag{63}$$

and $\bar{d} = d/l$ is the center-to-center spacing normalized on source streamwise length. The case $\bar{d} = 0$ indicates that the two sources overlap and the temperature is that of a single source at twice the dissipation. Interestingly, since the problem is linear, the temperature increase is twice $\Delta T_{2,o}$. The effect of the up-

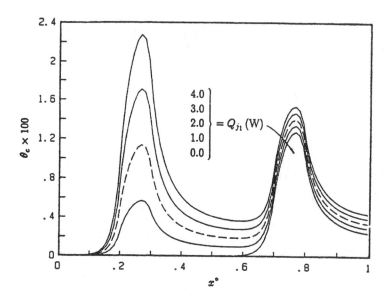

FIGURE 39. Spanwise centerline temperature for two sources, parametric on power dissipation of upstream source Q_1 for downstream source power $Q_2 = 2.0\,\text{W}$, $k_s = 2.0$ *W/m-K* [from Culham et al., 1991b].

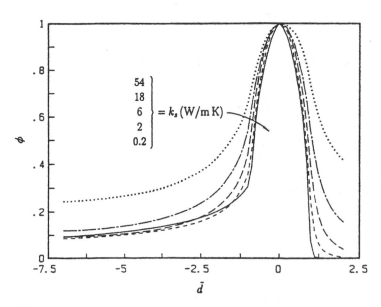

FIGURE 40. Effect on maximum temperature of downstream heat source due to location of upstream heat source, [from Culham et al., 1991b].

stream heat source is felt on the second source even when it is downstream strictly because of upwind board conduction, as can be seen by its increase with increasing conductivity.

UNIFORM FLOW

Ortega et al. [1993] proposed that the simplest model of the convective flow, one that assumes that the flow is a uniform parallel flow over the board, may be considered to be the most basic flow for the PCB cooling problem of Figure 34, and they used it to analyze both the heat transfer from the source and the behavior of the thermal wake downstream of the source. Sugavanam et al. [1995], for the two-dimensional problem, showed that the uniform flow solution provides the upper asymptotic limit for the board Nusselt number, since for a given inlet Reynolds number, it gives the highest near-wall streamwise velocity compared to either a boundary layer flow or a developed channel flow. Although the uniform flow model is not realistic in a strict sense, not satisfying, for example, the no-slip condition at the wall, it successfully predicts the behavior of the downstream thermal wake [Ortega et al., 1993] that has been experimentally observed in the turbulent flow over arrays of 3-D wall-mounted blocks simulating electronic components. The approach is discussed here primarily because it illustrates the use of full three-dimensional analytic kernel solutions that admit diffusion in the fluid in all directions.

The flow field is assumed to be a simple uniform flow parallel to the board with magnitude U leading to a degenerate form of the energy Equation 51

$$U\frac{\partial T}{\partial x} = \varepsilon\left(\frac{\partial^2 T}{\partial x^2} + \frac{\partial^2 T}{\partial y^2} + \frac{\partial^2 T}{\partial z^2}\right) \tag{64}$$

where ε is used as the symbol for the fluid effective thermal diffusivity so as not to be confused with the molecular diffusivity, α_f. If Equation 64 is used as a primitive model for a highly turbulent flow, the effective diffusivity ε may be interpreted as the sum of the molecular and the eddy diffusivities for heat [Kays and Crawford, 1993]

$$\varepsilon = \alpha_f + \varepsilon_H \tag{65}$$

leading to a definition for the "effective fluid conductivity"

$$k_e = \varepsilon\rho c \tag{66}$$

Henceforth, the model is referred to as the uniform flow effective diffusivity (UFED) model. Equation 64 may be nondimensionalized in the following way:

$$\frac{\partial\theta^*}{\partial X^*} = \left(\frac{\partial^2\theta^*}{\partial X^{*2}} + \frac{\partial^2\theta^*}{\partial Y^{*2}} + \frac{\partial^2\theta^*}{\partial Z^{*2}}\right) \tag{67}$$

where

$$\theta* = \frac{T - T_0}{\theta_{ref}} ; \quad X* = \frac{Ux}{\varepsilon} ; \quad Y* = \frac{Uy}{\varepsilon} ; \quad Z* = \frac{Uz}{\varepsilon} \tag{68}$$

θ_{ref} is a characteristic temperature for the problem. The fluid can be visualized as a bulk flow with homogeneous properties moving in the direction x over a plane surface; heat transfer occurs by diffusion in each principle direction and advection by the bulk flow. The kernel solution for a point source of strength Q moving at constant velocity over an infinite plane surface is available [Carslaw and Jaeger, 1959] and has previously been used in problems of moving sources of heat on the surface of conducting solids [Rosenthal, 1946] and in the analysis of film cooling [Eckert and Drake, 1972] by attaching the reference frame to the moving source. The solution is given by

$$\theta(x,y,z) = \left\{ \frac{Q}{2\pi\rho c\varepsilon R} \right\} \exp\left[-\frac{U}{2\varepsilon}(R - x) \right] \tag{69a}$$

where $R = \sqrt{x^2 + y^2 + z^2}$. In dimensionless form

$$\theta* = \frac{1}{2\pi R*} \exp\left\{ -\frac{1}{2}(R* - X*) \right\} \tag{69b}$$

where the temperature reference is given by

$$\theta_{ref} = \frac{Q}{k_e \, \varepsilon/U} \tag{69c}$$

The temperature field at $z = 0$ for the PCB cooling problem represents the solution for a point source on an adiabatic board with a uniform isotropic flow moving over it with a constant velocity U in the x direction. The point source solution is discontinuous at the origin but nevertheless may be used to generate solutions for sources of finite dimension [Ortega et al., 1993] of which the rectangular source is the most relevant for the printed circuit board cooling problem. Integrating Equation 69(a) over a rectangular source of dimensions $2b \times 2l$ in the streamwise and spanwise directions, respectively, and using the nomenclature of Figure 34 yields

$$\theta* = \frac{1}{2(2\pi)^{1/2}} \int_0^\infty \left\{ erf \frac{Y* + Pe_l}{2(2u)^{1/2}} - erf \frac{Y* - Pe_l}{2(2u)^{1/2}} \right\} \times$$

$$\left\{ erf \frac{X* + Pe_b - 2u}{2(2u)^{1/2}} - erf \frac{X* - Pe_b - 2u}{2(2u)^{1/2}} \right\} \frac{du}{u^{1/2}} \tag{70}$$

where

$$\theta_{ref} = \frac{q}{k_e \, U/\varepsilon} ; \quad Pe_b = \frac{Ub}{\varepsilon} , \quad Pe_l = \frac{Ul}{\varepsilon} \tag{71}$$

The source strength is now more appropriately indicated by q, heat flux on the surface of the source, assumed to be uniform.

Sample solutions are shown in Figure 41 for low Peclet number and in Figure 42 for high Peclet number. For illustration, the effective diffusivity has a value of 0.002 m²/s corresponding to the eddy diffusivity of turbulent flow of air over a flat plate. The surface plots clearly show the three-dimensional nature of the thermal wake brought about by diffusion in the x, y, and z directions and its response to increased Peclet number. At elevated Peclet number, the temperature over the source is nearly two-dimensional and one would expect the two-dimensional strip source solution to be accurate. Spanwise diffusion in the thermal wake is obvious at low *Pe* but relatively minor at the higher *Pe* as can be seen by comparison of Figures 41(b) and 42(b).

The UFED model was used to formulate the conjugate problem of Figure 34 by coupling it to a thin board model that was discretized with a Galerkin finite element method [Ramanathan, 1994 and Ramanathan and Ortega, 1996]. The analytical UFED model for the fluid domain provides a particularly simple coupling to the board FEM model allowing a noniterative procedure for obtaining the steady state temperature distribution over a rectangular board with a rectangular source. The analytical treatment of the board side follows that of Culham et al. [1991a] and is described by the thin board model given by Equation 57. The lateral edges and the backside of the board are assumed to be insulated. The domain of the board was discretized into bilinear rectangular elements and the Galerkin formulation for finite element analysis was applied to Equation 57. Bilinear shape functions were used for temperature within each element. The process resulted in the following matrix equation:

$$[KG][\theta] - [FG] + \left[\frac{qG}{k_s t}\right][q_f] - \left[\frac{QG}{k_s t}\right] = 0 \qquad (72)$$

where $[KG]$ is the global stiffness matrix, $[FG]$ is the matrix containing the edge boundary conditions (in this case a null matrix because of the insulated edge conditions), $[qG/k_s t]$ is the global coefficient matrix for the convective heat flux, and $[QG/k_s t]$ is the global matrix for the source term. $[q_f]$ is the single column matrix of unknown surface heat flux.

The point source solution for the flow, Equation 69(a), explicitly relates θ to q_f allowing a direct substitution into the board equations. In each discrete board element q_f is assumed to vary bilinearly, and the kernel solution is integrated over each element to compute the temperature distribution over the entire board due to the heat flux over each element. The temperature at every nodal point has contributions from all the elements on the surface. Discretizing of Equation 69(a) results in a matrix relation between θ and q_f

$$[\theta] = [A][q_f] \qquad (73)$$

where $[A]$ is the coefficient matrix for q_f, and inverting

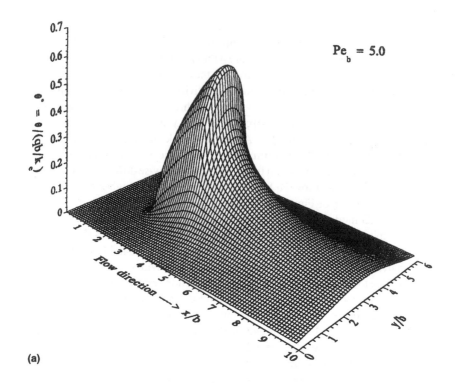

(a)

$$[q_f] = [B][\theta] \tag{74}$$

where

$$[B] = [A]^{-1} \tag{75}$$

Then substituting Equation 74 in Equation 72 and grouping the coefficients of the unknown temperatures into one matrix results in a system of equations in θ given by

$$\left[[KG] + \left[\frac{qG}{k_s t}\right][B]\right][\theta] = \left[\frac{QG}{k_s t}\right]. \tag{76}$$

The Gauss-Seidel iteration method was used to obtain the results presented here.

In order to observe the relevant dimensionless parameters, the board conduction equation, Equation 57, can be normalized using the definitions given in Equation 68 to give

$$Pe_b\left(\frac{k_s t}{k_e b}\right)\left(\frac{\partial^2 \theta^*}{\partial X^{*2}} + \frac{\partial^2 \theta^*}{\partial Y^{*2}}\right) - \frac{q_f}{q} + 1 = 0 \tag{77a}$$

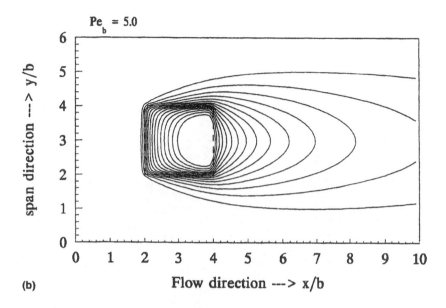

FIGURE 41. Dimensionless temperature rise of a square source on adiabatic surface, $Pe_b = 5.0$ $2b \times 2l = 2$ cm $\times$ 2 cm, $U = 1.0$ m/s, $\varepsilon = 0.002$ m²/s, $k_e = 2.362$ W/m-K; (a) surface plot, (b) surface isotherms.

where the temperature reference is

$$\theta_{ref} = \frac{q_s}{k_e\, U/\varepsilon} \tag{77b}$$

The source streamwise length scale b was introduced arbitrarily, hence Pe_b could just as well be Pe_l. Comparison of the flow side equation, Equation 67, and the solid side equation, (77a), reveals finally that the conjugate solution is expressible as

$$\theta^* = f\!\left(X^*, Y^*, Z^*\ ;\ Pe_b, Pe_l, \frac{k_e b}{k_s t}\right) \tag{78}$$

For discussion of representative results, the following base case was used: $t = 1$ mm and $2b \times 2l = 2$ cm $\times$ 2 cm. The board dimensions were $L_x \times L_y = 10$ cm $\times$ 6 cm for low Pe_b and 9 cm $\times$ 4 cm for high Pe_b. Nondimensional temperature fields are plotted for low Pe_b of 5.0 and high Pe_b of 500.0 each for conductivity ratios k_s/k_e of 1.0 and 10.0. The surface plots along with the corresponding contour plots are shown in Figures 43 and 44. Comparing the results for Pe_b of 5.0 and k_s/k_e of 1.0 with the adiabatic solution in Figure 41, it is clear that when board conduction is negligible, the spreading of the thermal wake is essentially due to the diffusion in the fluid. At higher Pe_b the convec-

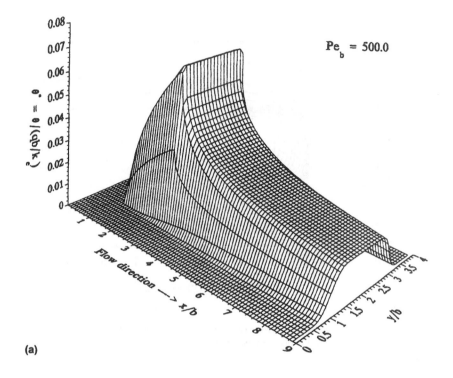

(a)

tion becomes dominant rendering the transport through conduction in both the solid and the fluid negligible. Comparison of the temperature fields on an adiabatic board for the same Pe_b (Figures 41 and 42) with those for k_s/k_e of 1.0 shows that at higher Pe_b the conduction in the solid is insignificant, and for the range of Pe_b considered, conductivity ratio of 1.0 does not introduce perceptible conduction effects. A tenfold increase in conductivity ratio enhanced the conduction heat transfer from the source at all Pe_b. Figures 43b, d and 44b, d show pronounced conduction and heat spreading in all directions from the source. At low Pe_b the conduction in the solid dominates, and the heat is spread more effectively on the board thus reducing the peak temperature as k_s/k_e increases. At higher Pe_b the convection dominates, and thus even a k_s/k_e of 10.0 does not yield dramatic heat spreading by board conduction as is evident in the contour plot, Figure 44(d). In the context of air cooling of PCBs, at high Pe_b the conduction effects are not significant enough to affect laterally displaced electronic components in the downstream wake; however, the strength of the thermal wake does emphasize the importance of thermal interaction of components along the flow direction by both board conduction and flow convection. Proper prediction of the thermal wake is critical.

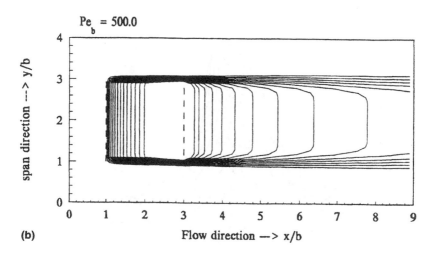

FIGURE 42. Dimensionless temperature rise of a square source on adiabatic surface, Pe_b = 500.0, $2b \times 2l$ = 2 cm × 2 cm, ε = 0.002 m²/s, k_e = 2.362 W/m-K; (a) surface plot, (b) surface isotherms.

A comparison of the thermal wake along the flow direction can be best made by considering the centerline temperatures. Figures 45 and 46 show the effect of conduction in the solid for Pe_b of 5.0 and 500.0, respectively. Each is also compared to the corresponding adiabatic board solution from the UFED model. It can be clearly seen from these plots that with increase in board conduction the thermal wake spreads preferably upstream since the convective wake downstream reduces the ability to remove heat downstream, hence reducing the conduction within the solid in that direction. Moreover, the plots show that for k_s/k_e ratio of 0.1 the board essentially behaves as adiabatic for all the Pe_b range analyzed. At higher Pe_b deviation of the thermal wake with board conduction, effects from the adiabatic wake becomes negligible. It should be noted that Pe_b of 5.0 are below any practical electronics cooling applications and that the conductivity of the fluid k_e is the effective conductivity that is defined with respect to the effective diffusivity (Equation 66). With respect to the molecular conductivity of the fluid, the conductivity ratio of 10.0 would yield solid conductivity close to that of epoxy boards used in electronic packaging. The plots of the centerline temperatures indicate that the thermal wake sufficiently downstream of the source is self-similar and as a first approximation may be computed with the adiabatic board results. The advantages of using such an analytical solution for the prediction of the thermal wake downstream of the source has been explored in detail by Ortega et al. [1993].

The average heat transfer coefficient or Nusselt number is a useful quantitative measure of the effectiveness of heat transfer from the source despite the fact that it is a defined quantity. In addition, it may be used in an uncoupled

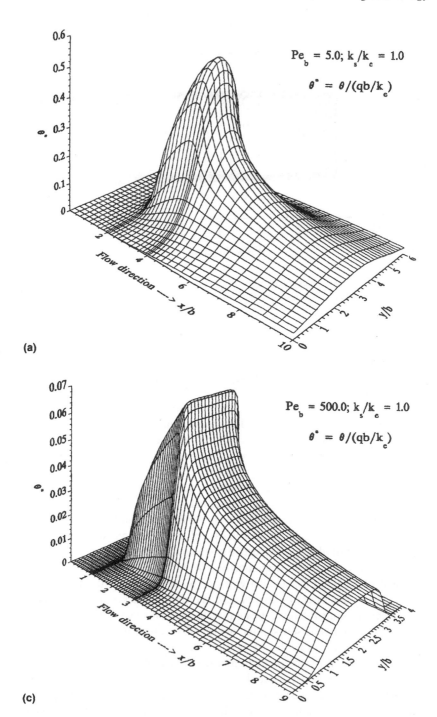

(a)

(c)

FIGURE 43. Nondimensional temperature rise due to a square heat source on conducting plate, surface plots: (a) $Pe_b = 5.0$, $k_s/k_e = 1.0$, (b) $Pe_b = 5.0$, $k_s/k_e = 10.0$, (c) $Pe_b = 500.0$, $k_s/k_e = 1.0$, (d) $Pe_b = 500.0$, $k_s/k_e = 10.0$.

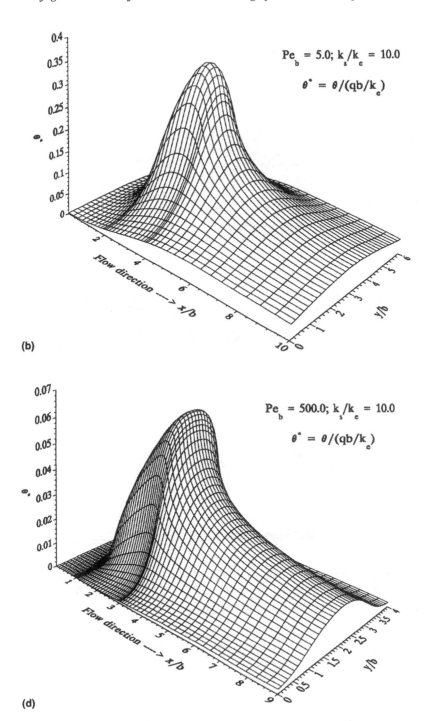

(b)

(d)

FIGURE 43. Continued

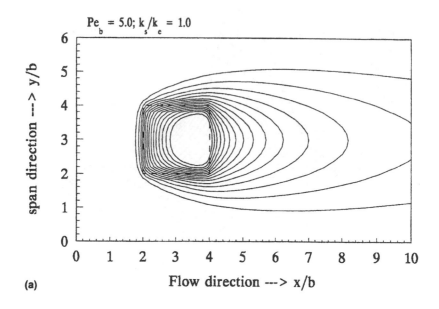

(a)

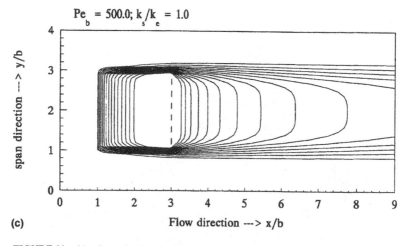

(c)

FIGURE 44. Nondimensional temperature rise due to a square heat source on conducting plate, surface isotherms: (a) $Pe_b = 5.0$, $k_s/k_e = 1.0$, (b) $Pe_b = 5.0$, $k_s/k_e = 10.0$, (c) $Pe_b = 500.0$, $k_s/k_e = 1.0$, (d) $Pe_b = 500.0$, $k_s/k_e = 10.0$

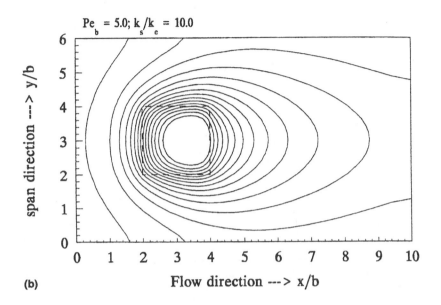

(b)

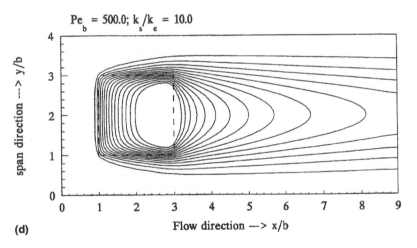

(d)

FIGURE 44. Continued

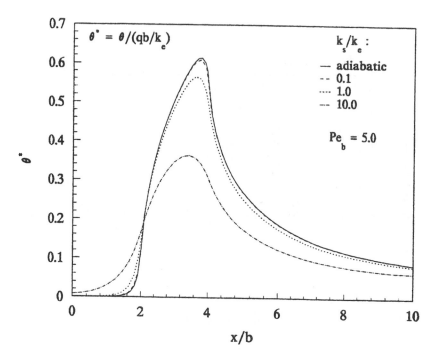

FIGURE 45.   Nondimensional spanwise centerline temperature rise, $l/b = 1.0$, $Pe_b = 5.0$.

analysis in such a way as to account for the degradation of convective heat transfer due to upwind heating through the board. Average Nusselt number is defined as

$$Nu_b = \frac{\overline{h}\,b}{k_e} \tag{79}$$

where $\overline{h}$ is the average heat transfer coefficient over the source.

The Nu_b were computed for various Pe_b ranging from 5.0 to 5000 and for $l/b = 0.5$, 1.0, and ∞. The case $l/b = \infty$ represents the infinitely wide 2-D strip of heat; the analytic solution for the adiabatic case is given in Ortega et al. [1993]. Nu_b plotted against the Pe_b is shown in Figure 47, logarithmic on both the axes. Curve fits were performed on the data for each aspect ratio. Above $Pe_b = 100.0$ there is no appreciable difference between the three aspect ratios, thus for the range of Pe_b encountered in practice, the Nu_b for a rectangular source of any aspect ratio l/b greater than 0.5 can be approximated as that for a square source. At low Pe_b the spanwise diffusion is significant, and hence the shape of the source affects the heat transfer from it. The magnitude of Nu_b varies inversely with l/b, being lowest for the 2-D case and highest for $l/d = 0.5$. A regression analysis on the square source data resulted in the following linear fit:

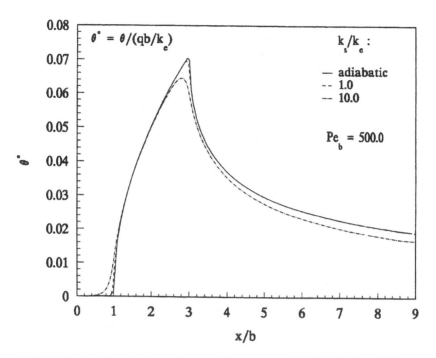

FIGURE 46. Nondimensional spanwise centerline temperature rise, $l/b = 1.0$, $Pe_b = 500.0$.

$$Nu_b = 1.24114 \, Pe_b^{0.498} \tag{80}$$

The standard deviation of the regression was 0.16. The error in using the square source correlation in predicting the strip source Nu_b at Pe_b of 5.0 and 5000.0 was 5.0% and 0.07%, respectively, and 2.7% and 0.8% for a rectangular source of aspect ratio 0.5 at Pe_b of 5.0. An interesting note about the exponent on the Peclet number in Equation 80 is that an analysis of the two-dimensional problem with negligible streamwise diffusion shows the exponent on Peclet number to be exactly 0.5.

To determine the response of the heat transfer over the source to variations in Pe_b at various board conductivities, Nu_b for various Pe_b ranging from 5.0 to 1000.0 and for k_s/k_e of 0.1, 1.0 and 10.0 were computed for the square source. After extensive analysis, two correlations were developed [Ramanathan, 1994]. In the first case, a correlation was developed in such a way that average Nusselt number tended to the adiabatic board limit as conductivity ratio k_s/k_f approached zero. The result was

$$\frac{Nu_b}{Nu_{b(ad)}} = 1 - 0.203 \, Pe_b^{-0.174} \left(\frac{k_s}{k_e}\right)^{0.334489 \, Pe_b^{0.0368}} \tag{81}$$

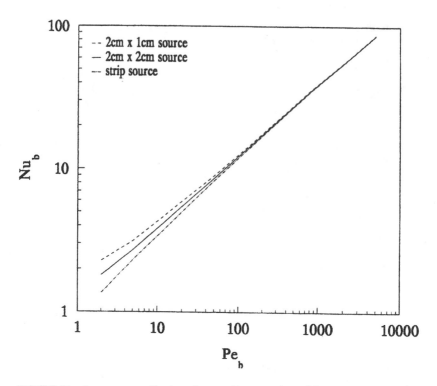

FIGURE 47.   Average source Nu_b dependence on Pe_d, comparison of three source aspect ratios on an adiabatic board.

and is useful for $0.0 \leq k_s/k_f \leq 10.0$. An expression that does not tend to the correct adiabatic limit but is simpler to use is

$$Nu_b = 0.943349(k_s/k_e)^{-0.05} \, Pe_b^{0.53} \qquad (82)$$

applicable for $Pe_b = 5.0$ to 1000.0 and $k_s/k_e = 0.1$ to 10.0. The goodness of the correlation is tested in Figure 48 for all the parameter variations attempted. These correlations offer a way to characterize the conjugate heat transfer from a square source of heat, and in particular, to account for the effects of board conduction in an uncoupled analysis.

CONCLUSION

This chapter has presented analysis of and results from a limited class of conjugate subproblems of the printed circuit board cooling problem, problems in which the flow is sufficiently simple so that the conjugate mechanisms can be emphasized without being obscured by difficult fluid flow physics or numerical issues that invariably arise in CFD analysis of more difficult geome-

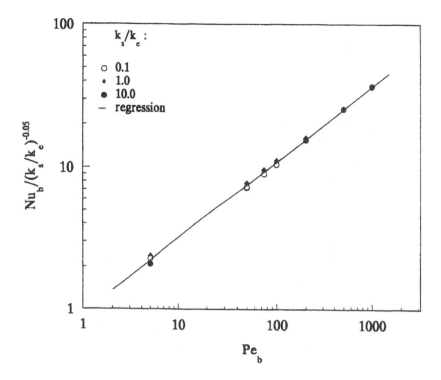

FIGURE 48. Correlation of source average Nusselt number for various board conductivities.

tries. It has become increasingly possible to apply commercially available CFD codes for the solution of fully conjugate problems of the type discussed in this chapter but with greater complexity in both the geometry and in the thermal boundary conditions. Readers are referred to the study of Choi et al. [1994] and the references reviewed in that study in which the geometry of Figure 49 has been extensively analyzed using a commercially available CFD tool. The geometry is a two-dimensional representation of a printed circuit board with a somewhat detailed model of the electronic package modules mounted on the board surfaces. The solution approach is nearly identical to that of Sugavanam et al. [1995] described in this chapter. Because such problems introduce a very large number of parameters, it is difficult to extract broadly applicable results from such studies. Nevertheless, the ability to successfully compute such geometries is by now readily available in commercial form and is limited only by computing resources and computer resourcefulness.

It is worth remembering that a thermal analyst's toolbox should contain many vertical compartments; we earlier referred to them as hierarchies. The uppermost, most accessible compartment should contain simple, easy to use tools such as design rules of thumb that can be developed on the back of an envelope.

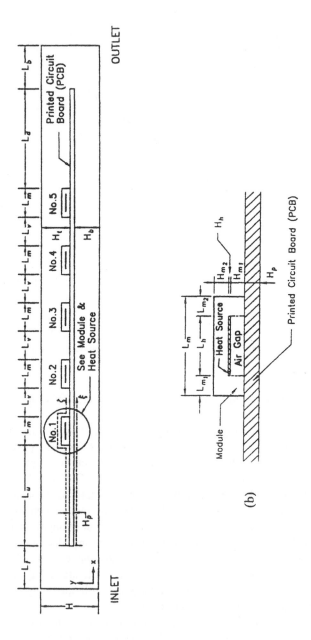

FIGURE 49. Electronic components mounted on a conducting printed circuit board, two-dimensional representation [after Choi et al., 1994].

When the problem resists solution with the tools in the uppermost compartment, the tools in the next lower compartments should be pulled out. At the lowermost compartment, the analyst should keep the biggest hammers and most sophisticated tools. The successful analyst will probably not have to resort to that lowest compartment very often. It is the author's hope that this chapter, having demonstrated various ways to approach the PCB cooling problem by use of tools from different compartment levels (but primarily from the mid-level ones), will prompt users to question their choice of tools and expand their toolbox.

ACKNOWLEDGMENTS

Much of the work presented would not have been possible without the support of my sponsors. With gratitude I acknowledge the support over the past five years of the National Science Foundation, the Semiconductor Research Corporation, the IBM Corporation, the Intel Corporation, and Advanced Micro Devices. My colleagues Drs. Chris Choi of the University of Arizona, Sung Jin Kim of the IBM Corporation, and Cho-lik Chan of the University of Arizona have taught me much about numerical techniques in conjugate heat transfer, and I thank them for their friendship and technical contributions. Darvin Edwards of Texas Instruments and Tom Tarter of AMD have been invaluable mentors through the sponsorship of the SRC. It has been a pleasure to explore these issues with my assistants Humayun Kabir, Shankar Ramanathan, Ramesh Sugavanam, and Uli Wirth who have contributed immeasurably to the technical and not so technical aspects of my work.

REFERENCES

Anderson, A.M. and Moffat, R.J. 1990a. *Convective heat transfer from arrays of modules with non-uniform heating: experiments and models*, Report HMT-43, Thermosciences Div., Mech. Eng. Dept., Stanford University, Stanford, CA.

Anderson, A.M. and Moffat, R.J. 1990b. A new type of heat transfer correlation for air cooling of regular arrays of electronic components, ASME HTD - 153. *Thermal Modeling of Electronic Systems and Devices*, ASME, New York, 27–40.

Brebbia, C.A., Telles, J.C.F., and Wrobel, L.C. 1984. *Boundary Element Techniques: Theory and Application in Engineering*, Springer-Verlag, New York.

Carslaw, H.S. and Jaegar, J.C. 1959. *Conduction of Heat in Solids*, 2nd ed, Clarendon Press, Oxford.

Choi, C. Y., Kim, S. J., and Ortega, A. 1994. Effects of Substrate conductivity on convective cooling of electronic components, *ASME J. Elec. Pakag.*, 116:198–245.

Culham, J.R., Lemcyzk, T.F., Lee, S., and Yovanovich, M.M. 1991a. META-a conjugate heat transfer model for air cooling of circuit boards with arbitrarily located heat sources, ASME HTD-171, *Heat Transfer in Electronic Equipment*, A. Ortega, D. Agonafer, and B. Webb, Eds., ASME, New York, 117–126.

Culham, J.R. and Yovanovich, M.M. 1991b. The effect of common design parameters on the thermal performance of microelectronic equipment: II. Forced convection, ASME HTD-171,

Heat Transfer in Electronic Equipment, A. Ortega, D. Agonafer, and B. Webb, Eds., ASME, New York, 55–62.

Eckert, E.R.G. 1950. *Introduction to the Transfer of Heat and Mass*, McGraw-Hill, New York.

Eckert, E.R.G. and Drake, R.M. 1972. *Analysis of Heat and Mass Transfer*, McGraw-Hill, New York.

Gorski, M.A. and Plumb, O.A. 1990. Conjugate heat transfer from a finite strip heat source in a plane wall. ASME HTD-129, *AIAA-ASME Thermophysics and Heat Transfer Conference*, Seattle, ASME, New York, 47–53.

Gorski, M.A. and Plumb, O.A. 1992. Conjugate heat transfer from an isolated heat source in a plane wall. ASME HTD-210, *Fundamentals of Forced Convection Heat Transfer*, M.A. Ebadian and P.H. Oosthuizen, ASME, New York, 99–105.

Gorski, M.A., 1992. *On the Heat Transfer from a Small, Discrete Heat Source*, Ph.D. thesis, Washington State University, Pullman.

Incropera, F.P., Kerby, J.S., Moffat, D.F., and Ramadhyani, S. 1986. Convection heat transfer from discrete heat sources in a rectangular channel, *Int. J. Heat Mass Transfer*, 29:1051–1058.

JEDEC. 1994. Methodology for the thermal measurement of component packages (single semiconductor devices), JC15.1 subcommittee test proposal, JEDEC.

Jensen, R.H., Andrejack, G.A., Button, D.P., and Bydal, B.A. 1989. Comparative thermal performance of various substrate materials in a simple packaging application: actual versus predicted in *Proc. 39th IEEE Electronic Components Conference*, IEEE, New York, 572–576.

Kabir, H., Ortega, A., and Chan, C.L. 1995. A boundary element formulation of the conjugate heat transfer from a convectively cooled discrete heat source mounted on a conductive substrate. *IEEE Transfer Comp., Packag. Mfg. Technol-Part A.*, 18(1):108–116.

Kays, W.M. and Crawford, M.E. 1993. *Convective Heat and Mass Transfer*, 3rd ed., McGraw-Hill, New York.

Krause, W.B., Klasi, M.L. and Addala, S.R. 1989. Conjugate heat transfer from isothermal surface mounted heat sources embedded in a large substrate. ASME HTD, 111, *Heat Transfer in Electronics-1989*, 1989 Heat Transfer Conference, 161–168.

Lee, C.C. and Palisoc, A.L. 1990. *Topics in Boundary Element Research: Electrical Engineering Application*, C.A. Brebbia, Ed. Vol. 7, Springer-Verlag, New York.

Lemczyk, T.F., Mack, B.F., Culham, J.R., and Yovanovich, M.M. 1991. PCB trace thermal analysis and effective conductivty in *Proc. 7th Annual IEEE Semiconductor Thermal Meas. and Manag. Symp.* (SEMITHERM), IEEE, New York, 15–22.

McCuen, P.A. 1962. Heat transfer with laminar and turbulent flow between parallel planes with constant and variable wall temperature and heat flux, Ph.D. thesis, Stanford University, Stanford.

Moffat, R.J. and Ortega, A. 1988. Direct air cooling of electronic components, in *Advances in Thermal Modeling of Electronic Components and Systems*, Vol I, A. Bar-Cohen and A.D. Kraus, Eds., Hemisphere, New York.

Ortega, A., Ramanathan, S., Chicci, J.D., and Prince, J.L. 1993. Thermal wake models for forced air cooling of electronic components in *Proc. 9th Annu. IEEE Semiconductor Thermal Meas. and Manag. Symp.* (SEMITHERM). IEEE 93CH3226-8, New York, 63–74.

Ortega, A., Wirth, U., and Kim, S.J. 1994. Conjugate forced convection from a discrete heat source on a plane conducting surface: a benchmark experiment, ASME HTD-292, *Heat Transfer in Electron. Sys.*, ASME, New York, 25–36.

Patankar, S.V. 1980. *Numerical Heat Transfer and Fluid Flow*, Hemisphere, New York.

Peterson, G.P. and Ortega, A. 1990. Thermal control of electronic equipment and devices, in *Advances in Heat Transfer*, Vol. 20, J.P. Hartnett and T.F. Irvine, Jr., Eds, Academic Press, New York.

Ramadhyani, S., Moffatt, D.F., and Incropera F.P. 1985. Conjugate heat transfer from small isothermal heat sources embedded in a large substrate, *Int. J. Heat Mass Transfer*, 28:1945–1952.

Ramanathan, S. 1994. Uniform Flow Effective Diffusivity Model for Convective Cooling of Electronic Components, Master's thesis, University of Arizona, Tucson.

Ramanathan, S. and Ortega, A. 1995. A uniform flow effective diffusivity approach for conjugate forced convection from a discrete rectangular source on a plane conducting surface, Procs. 1996 IEEE ITHERM Symposium.

Rosenthal, D. 1946. The theory of moving sources of heat and its application to metal treatment, *ASME J. Heat Transfer*, 849–866.

Rubesin, M.W. 1945. An Analytical Investigation of the Heat Transfer Between a Fluid and a Flat Plate Parallel to the Direction of Flow Having a Stepwise Discontinuous Surface Temperature, Master's thesis, University of California at Berkeley, Berkeley.

Rubesin, M.W. 1951. The effect of an arbitrary surface temperature variation along a flat plate on the convective heat transfer in an incompressible turbulent boundary layer, *NACA TN*, 2345, NACA.

Seban, R.A. 1950. Calculation method for two-dimensional boundary layers with arbitrary free stream velocity variation and arbitrary wall temperature variation, Inst. of Engrg. Research, Univ. of California at Berkeley, Berkeley.

Tribus, M. and Klein, J. 1952. Forced convection from non-isothermal surfaces, in *Proc. Heat Transfer Symp.*, University of Michigan, 211–235.

Sugavanam, R. 1994. Numerical Investigation of Conjugate Heat Transfer from Flush Heat Sources on a Conductive Board in Laminar Channel Flow, Master's thesis, University of Arizona, Tucson.

Sugavanam, R., Ortega, A. and Choi, C.Y. 1995. A numerical investigation of conjugate heat transfer from a flush heat source on a conductive board in laminar channel flow, *Int. J. Heat and Mass Transfer*, 1995, 38:2969–2984.

Tummala, R.R. and E.J. Rymaszewski, Eds. 1989. *Microelectronics Packaging Handbook*, Van Nostrand Reinhold, New York.

Wirth, U. 1994. Experimental Investigation of Conjugate Heat Transfer from a Flush Mounted Heat Source on a Conductive Substrate in Laminar and Turbulent Flows, Master's thesis, University of Arizona, Tucson.

FOR FURTHER INFORMATION

In addition to the references cited in the text, readers are referred to these additional references for further information about conjugate heat transfer as related to electronics cooling issues.

Brosh, A., Degani, D. and Zalmanovich, S. 1992. Conjugated heat transfer in a laminar boundary layer with heat source at the wall, *ASME J. Heat Transfer*, 11:709–724.

Davalath, J. and Bayazitoglu, Y. 1987. Forced convection cooling across rectangular blocks, *ASME J. Heat Transfer*, 109:321–328.

Kim, S.H. and Anand, N.K. 1994. Laminar developing flow and heat transfer between a series of parallel plates with surface mounted discrete heat sources, *Int. J. Heat Mass Transfer*, 37:2231–2244.

Kim, S.H. and Anand, N.K. 1994. Turbulent heat transfer between a series of parallel plates with surface-mounted discrete heat sources, *ASME J. Heat Transfer*, 116:577–587.

Nigen, J.S. and Amon, C.H. 1994. Time-dependent characteristics of conjugate heat transfer characteristics of self-sustained oscillatory flows in a grooved channel, *ASME J. Fluids Eng.*, in press.

Zebib, A. and Wo, Y.K. 1989. A two dimensional conjugate heat transfer model for forced air cooling of an electronic device, *ASME J. Electron. Packag.*, 9:11–16.

Chapter 5

ENHANCED AIR COOLING
OF ELECTRONIC EQUIPMENT

Suresh V. Garimella

CONTENTS

0-8493-9447-3/96/$0.00+$.50
© 1996 by CRC Press, Inc.

INTRODUCTION

In view of the simplicity of implementation and cost effectiveness of using air as the cooling medium for electronic equipment, it is highly desirable to explore means for enhancing the cooling capabilities of air before recourse is made to more aggressive cooling techniques such as using liquids in free or forced convection, in a duct-flow, an impingement, or a spray-cooling arrangement. Since unassisted (unenhanced) air cooling is not adequate for many applications, a thorough investigation of enhancement techniques is crucial. This chapter covers various enhancement techniques including the use of geometric adjustments such as staggering chip-arrays; surface enhancements in the form of a large variety of fins; and passive flow-modulation devices such as turbulators, vortex generators, barriers, displaced promoters, two-dimensional ribs, and three-dimensional roughness elements. Subsequently, jet impingement from single and multiple arrays of round and slot jets will be discussed as well as microchannel and heat-pipe heat sinks, porous metallic matrix inserts, improved substrates, and hybrid techniques.

ENHANCEMENT STRATEGIES

Air-cooling techniques may be broadly classified, based on driving force, into free-convection and forced-convection techniques. This chapter is primarily concerned with forced convection. Detailed discussions of free convection air cooling and its enhancement are beyond the scope of this chapter. For information on these topics, the reader is referred to reviews of the literature available in Aihara [1991] and Moffat and Ortega [1989].

A large number of choices exist for the enhancement of forced-convection cooling of electronic equipment. The following strategies may be identified for the design and implementation of any enhancement technique:

- Improvement of convective heat transfer coefficient: Improvement can be accomplished by the choice of a fluid with superior thermophysical properties or by increasing flow velocities. Since this chapter deals with air cooling, the former option cannot be considered further. The latter option, on the other hand, may come at a disproportionate cost in terms of the required pumping power and is often not a preferred option. It is worth noting in this connection that the so-called Mouromtseff number may be used to evaluate the thermal performance of a coolant in a given configuration [Chu et al., 1970]; this number simply recasts the appropriate forced-convection correlation to reflect the dependence of the Nusselt number on the thermophysical properties included in the nondimensional parameters of the correlation. The Mouromtseff number for different fluids for turbulent flow through smooth tubes is shown in Figure 1 as a

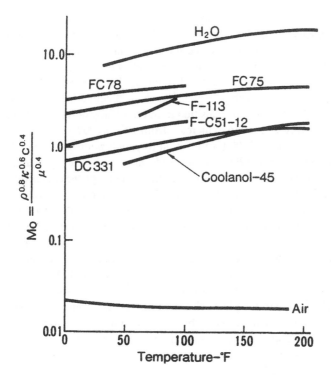

FIGURE 1. Mouromtseff number for fully developed turbulent flow in smooth tubes (Chu et al. [1970], by permission).

function of temperature; the higher the value of this number, the better is the heat transfer performance of the fluid.

- Increase of surface area available for heat transfer: This has traditionally been accomplished by the use of enhanced surfaces (fins), and a large body of literature exists on their design and implementation.
- Flow modulation: This strategy aims to increase mixing in the flow using devices such as vortex generators, turbulators, and swirl flow devices. Increases due to flow modulation in the heat transfer coefficient of as much as 100% have been reported for turbulent flows.

These three strategies may often not be entirely distinct; for instance, flow modulation is essentially a means for increasing heat transfer coefficients by selectively increasing the local velocity or shear in the flow near the surface to be cooled. Similarly, arrangements such as louvered fins increase both the heat transfer coefficient as well as the surface area [Webb, 1987].

Different strategies are typically effective in the different flow regimes—laminar, transitional, and turbulent. Flow-regime transition has been well characterized in the literature for smooth ducts and pipes. However, when flow channels carrying electronic components are considered, the definition of laminar to turbulent transition is less obvious. This transition was investigated using flow visualizations and turbulence measurements in water by Garimella and Eibeck [1992]. The onset of transition was shown to be a function of channel height as well as the chip layout and geometry. For instance, as channel height increased from 1 to 3.6 chip heights, the channel height-based Reynolds number (UH/υ) for the onset of transition in interior (fully developed flow) regions of the chip array was found to increase from 700 to 1900. This compares to the smooth-pipe value of 2400 for transition Reynolds number based on hydraulic diameter, or roughly 1200 for Reynolds number based on channel height. Garimella and Eibeck [1992] also suggested that though the chip height was not a variable in their study, it appeared that a Reynolds number based on chip height appeared to be the correct predictor for transition, with transition occurring for UB/υ > 550 at a chip spacing of about 2 chip heights; this threshold was lowered to 400 at larger spacings of 6 chip heights. Igarashi and Takasaki [1992] proposed a similar threshold of UB/υ > 900 for transition in air. Lower limits of transition in a rectangular duct are tabulated in Bhatti and Shah [1987], where the critical Reynolds number is listed as a function of duct aspect ratio and varies from 3100 for a parallel-plates channel to 2200 for a square duct. The critical Reynolds number is also affected by the conditions at the entrance to the channel—an abrupt entrance is more susceptible to transition than a smooth entrance.

Individual techniques to achieve these enhancement objectives are discussed in the rest of this chapter.

ENHANCEMENT TECHNIQUES

INHERENT ENHANCEMENT AT ELECTRONIC COMPONENT SURFACES

It is important to recognize that electronic components have naturally enhanced heat transfer (due to enhanced hA products) when compared to predictions based on thermally ordered conditions (smooth surface, vibration-free analyses). In addition to the beneficial flow-field disruption caused by virtue of their protrusion, this natural enhancement results from the surface roughness of the circuit boards, vibrations in electrical equipment, and electric fields in power equipment [Kraus and Bar-Cohen, 1983]. The surface roughness on a printed circuit board has been shown to lead to an increase in the heat transfer coefficient relative to a smooth surface by up to a factor of 10 [Wenthen, 1977].

ENHANCEMENT BY CONTROLLING GEOMETRIC LAYOUT

Significant enhancements in heat transfer can be realized by controlling the geometric layout of the chips and the assembly of the boards. Staggering the chips on a board has been shown in several studies to increase the heat transfer coefficient relative to an inline configuration for a given inlet velocity and channel height. In air-cooling experiments, Hollworth and Fuller [1987] obtained as much as a 50% increase in heat transfer; in water, a 10 to 40% enhancement was seen to be accompanied by a 20 to 110% increase in pressure drop [Garimella and Eibeck, 1991c]. Figure 2 shows the heat transfer coefficients for inline and staggered arrangements of an array of chips with water as coolant [Garimella, 1991]. The flow visualizations in Figure 3 (water with hydrogen-bubble tracers and laser-sheet illumination) demonstrate the superior mixing obtained by staggering the chips in an array. With fan power held constant, Ashiwake et al. [1983] obtained a 70% drop in cooling-air temperature by staggering the chips of an inline array.

Even in inline arrays, the heat transfer coefficient can be increased by increasing either the streamwise spacing between chips (35–40%) or the spanwise spacing (15%) as shown in Garimella and Eibeck [1990, 1991b]. General recommendations about component orientation and the use of mixed compo-

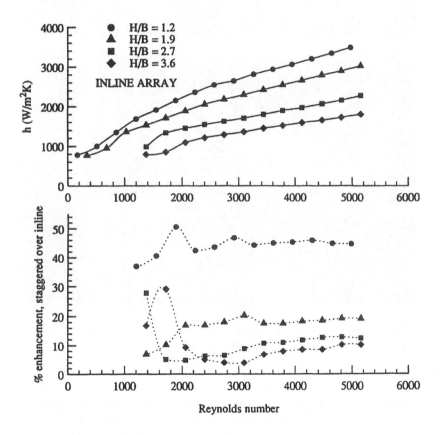

FIGURE 2.   Heat transfer coefficients in water for an inline array of chips, and the percent enhancement obtained by fully staggering the chips (by one chip width); the Reynolds number (Re_H) is based on the channel height, and the chip spacing is equal to the chip width and length.

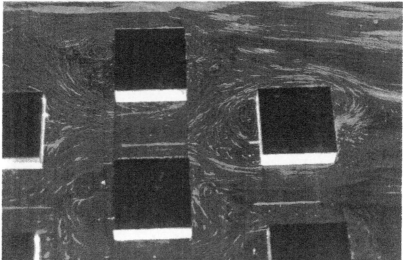

FIGURE 3. Flow patterns visualized with hydrogen bubbles in water for inline and staggered arrays of chips; flow is from right to left (H/B = 3.6, Re$_H$ = 3450).

nent sizes were drawn by Azar and Russell [1991] based on flow patterns visualized in water. There are also indications that chips in the horizontal orientation have lower thermal resistances than those in the vertical orientation, even in forced convection due to buoyancy effects; the vertical orientation also results in greater asymmetry in component temperature distribution [Azar et al., 1989]. This effect would, however, be swamped at sufficiently large Reynolds numbers.

When possible, chips in a horizontal channel should be mounted on the bottom plate (facing upwards), so as to benefit from the enhancement effect from free convection; in contrast, a heated top-plate configuration displays a thermally stratified behavior that suppresses the effect of buoyancy. The orientation of the chips is important at the lower—laminar and weakly turbulent—range of Reynolds numbers. For instance, Osborne and Incropera [1985] showed that the heat transfer coefficients for the heated bottom plate in a horizontal parallel-plates duct could be as much as 50% larger than for the top plate at a weakly turbulent Reynolds number based on channel height of 5200.

Another system parameter of control may be high free-stream turbulence. This information is included under controlling geometric layout, since it may be possible to enhance the turbulence level of the incoming coolant stream either by appropriate choices in manifold design, or by placing turbulence-promoting peripheral devices upstream from the chips. Simonich and Bradshaw [1978] reported significant increases in Stanton number in response to high free-stream turbulence. Pedisius et al. [1979] found increases in heat transfer of 10 to 15% with turbulence intensities up to 8%; upon further increasing the turbulence intensity to 14%, however, only a 20% increase in heat transfer was observed. Blair [1983a, b] reported heat transfer enhancements of as much as 20% for a grid-generated free stream turbulence of 7%. While there have been different explanations in the literature of the precise effect of free-stream turbulence on heat transfer, it appears that free-stream turbulence not only causes transition to occur earlier (at lower Reynolds numbers) but also has an effect on the fully turbulent boundary layer [Blair, 1983a, b]. Maciejewski and Moffat [1989] reported increases in heat transfer of 1.8 to 4 times relative to that predicted using accepted correlations for turbulent boundary layers, due to very high free-stream turbulence intensities of 20 to 60%.

ENHANCED SURFACES

A vast amount of literature exists on the use of extended surfaces for heat transfer enhancement. Reported studies have ranged from analytical solutions for the simpler fin configurations to detailed conduction and conjugate heat transfer computations and to experimental databases for a large variety of extended surfaces. Detailed discussions are widely available (for instance, see Bergles et al. [1983, 1991]; Webb [1987, 1994]) and this material will not be covered in the present chapter.

"LARGE" ROUGHNESS ELEMENTS (2-D RIBS AND 3-D PROTRUSIONS)

As a general rule, obstructions are less effective in laminar flow than in turbulent flow. The laminar sublayer in turbulent flow is a lot more fragile and thus more amenable to being disrupted. However, obstructions may help to trigger transition at lower Reynolds numbers in an otherwise smooth channel. Thus in a narrow range of flow rates at the upper end of the laminar regime, obstructions can cause sharp increases in heat transfer.

Electronic components themselves often act as protrusions that enhance mixing in the flow [Garimella and Eibeck, 1991c]. While the layout of the components on the board is mostly governed by nonthermal hardware considerations, the strategic placement of additional barriers could still be used to significant advantage.

Numerous studies of two-dimensional rectangular ribs have appeared in the literature. Han et al. [1978] studied the effect of rib geometry (shape, angle of attack, and pitch-to-height ratio) on friction factor and Stanton number for fully developed turbulent air flow. Ribs were placed on both walls with the rib height being less than 15% of the channel height. Both the ribs and the interrib spaces were heated. The ribs were treated as classical roughness and a maximum in both friction factor and Stanton number was found to occur at a rib spacing to height ratio of 9. This spacing was reported to correspond roughly to the situation where separated flow from a rib reattaches just prior to the next rib. Interestingly the cross-sectional shape of the rib had only a modest effect on heat transfer, while the friction factor was affected markedly.

Velocity and turbulence characteristics in the vicinity of a two-dimensional rib were presented by Crabb et al. [1977]. The reattachment point was found to be approximately 12-rib-heights downstream of the leading edge. The height and streamwise length of the rib were shown to affect the trajectory of the streamline dividing forward and backward flow (zero-velocity streamline), and hence, the reattachment length. This was proposed as a reason for the observed difference in reattachment lengths behind a rib, a fence, and a backward-facing step. Significant fluctuating energy at a discrete frequency was found to be present in the vicinity of the shear layer above the rib.

The reattachment length behind a two-dimensional obstacle (backward-facing step) increases markedly with Reynolds number in the laminar regime, with a sudden drop at transition, and a gradual increase with Reynolds number in the turbulent regime [Durst et al., 1986; Eaton and Johnston, 1981; Tropea and Gackstatter, 1985].

Three-dimensional obstacles behave very differently from their two-dimensional counterparts. The primary difference comes from the behavior of the flow field in terms of the three-dimensional separation and reattachment. For instance the reattachment length is on the order of 1 to 4 obstacle heights in the three-dimensional case as opposed to 11 to 15 rib heights for two-dimensional ribs, although the trends of variation with Reynolds number are similar for the two kinds of obstructions. The three-dimensional separation and reattachment process is more vigorous due to the reattachment being from three sides (as opposed to solely from the top for a rib) and has a greater impact on heat transfer. Maximum turbulent stresses occur near the top of the obstacle. The wake behind both cylindrical and cubical obstructions disappears almost completely by a distance of about five heights downstream, and no velocity de-

fect remains [Castro and Robins, 1977; Lim and Lewkowicz, 1986]. As
Reynolds number increases, the lateral spread of the fluid dynamic and thermal wake decreases [Lim and Lewkowicz, 1986; Morris et al., 1995]. Greater
turbulence intensities in the upstream flow have also been found to reduce the
size of the recirculation region downstream of an obstruction. The horseshoe
vortex shed from a three-dimensional obstacle mounted on a wall may also have
an impact on heat transfer enhancement; this is discussed in a later section.

Garimella and Schlitz [1995] recently studied heat transfer enhancement in
a rectangular duct with very small height to width ratio (0.017) simulating
stacked printed circuit boards carrying computer chips using two- and three-
dimensional mixing devices. Their experiments were performed over a wide
range of transitional and turbulent Reynolds numbers in water and FC-77 (a
perfluorinated dielectric fluid). Heat transfer and pressure drop results were
presented for a discrete, flush heat source, a protruding heat source, and for an
array of protrusions. Heat transfer enhancement obtained by these means was
measured, as well as that obtained by the introduction of transverse ribs on the
opposite wall. The greatest enhancement of 100% relative to the flush heat
source was obtained for the array of roughness elements used in conjunction
with a rib on the wall opposite the heated element. Results from this work are
shown in Figure 4, where the enhancement is defined with respect to a flush-
mounted chip.

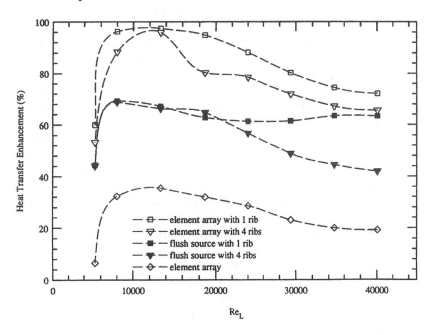

FIGURE 4. Extent of heat transfer enhancement (FC-77, relative to flush chip) of the protruding chips and with two-dimensional ribs placed on the opposite wall [Garimella and Schlitz, 1995].

CLASSICAL ROUGHNESS

Roughness elements, either random in nature or in a repeated-rib configuration, have been extensively studied as a technique for heat transfer enhancement in channel flows. The roughness elements considered are typically much smaller than the ribs and obstructions of the preceding section—at least an order of magnitude less than the boundary layer thickness, and in tubes, one or two orders of magnitude less than the tube diameter. Nakayama [1982] reviewed the influence of roughness Reynolds number, Prandtl number, and geometrical parameters on the momentum and heat transfer roughness functions, for both granular, three-dimensional surface roughness and for repeated-rib roughness. It was concluded from comparing a large number of studies in the literature that the relative merits of roughening a surface are large when the roughness Reynolds number is small; granular (3-D) roughness provided more favorable results than rib (2-D) roughness. The heat transfer enhancement is also larger for higher Prandtl-number fluids. It should be pointed out that in implementing roughness as a heat transfer enhancement technique variable physical properties have a more pronounced effect on heat transfer in rough passages than in smooth passages [Wassel and Mills, 1979]. Since fairly large temperature variations are experienced in electronic applications, care should be taken to account for variable-property effects.

Heat transfer and friction in tubes with rib-roughness (0.01 to 0.04 times tube diameter in height) were modeled as classical roughness by Webb et al. [1971]. The correlation for friction factor was based on the law-of-the-wall similarity used by Nikuradse [1933] for sand-grain roughness, while the heat transfer correlation was based on a heat and momentum transfer analogy (also used by Dipprey and Sabersky [1963]). The authors argued that this method of correlation could be used for geometrically similar forms of arbitrary roughness.

Perry et al. [1969] identified two types of two-dimensional roughness elements: *d-type* associated with narrow cavities where the roughness function (which describes the deviation of the log-law velocity profile from the smooth-wall behavior) depends on duct diameter, and *k-type* with larger spacing between roughness elements where the roughness function depends on the roughness height. It is this latter k-type roughness that, when installed upstream from electronic components, could be used for enhancing heat transfer. This type of roughness is characterized by an interelement spacing greater than about three times the roughness height, resulting in the flow between roughness elements interacting strongly with the core flow, with eddies shed into the core. Townes and Sabersky [1966] also observed a periodic instability when the cavities between roughness elements were shallow (small element height, large spacing); in this case, the vortex formed in the cavity broke down and interacted with the core flow. These instabilities result in higher heat transfer rates.

Kader and Yaglom [1977] made the first clear distinction between two- and three-dimensional roughness. The heat transfer dependence on roughness Reynolds number derived in their model was different for the two kinds of roughness. A similar difference in heat transfer dependence was found by Garratt and Hicks [1973] and Webb et al. [1971]. The influence of the roughness-element geometry on the heat transfer and friction characteristics has been investigated by several researchers [Baumann and Rehme, 1975; Dalle Donne and Meyer, 1977; Hall, 1962; Maubach, 1972; Meyer, 1980].

VORTEX GENERATORS AND BARRIERS

Significant work has been done for decades on the use of vortex generators to enhance heat transfer and to control flows in different applications. Figure 5 illustrates the principle of operation of a half-delta-wing vortex generator and the longitudinal vortex formed; a pair of counter-rotating vortex generators are also shown.

Fundamental studies of longitudinal vortices imbedded in turbulent boundary layers were carried out by Eibeck and Eaton [1987] and Wroblewski and Eibeck [1991] among others. Wroblewski and Eibeck showed that the vortex interaction with the turbulent boundary layer enhanced the heat transfer to a greater extent than did the momentum transport. Heat transfer coefficients and friction factors were obtained as a function of Reynolds number and angle of attack using rectangular, delta-wing, and half-delta-wing vortex generators by Fiebig et al. [1986]. Delta wings were found to yield the greatest enhancement, with the ratio of Colburn factor to friction factor being the highest at small angles of attack (10 deg). Russell et al. [1982] showed that a value for this ratio of more than 0.5 could be achieved with counter-rotating vortex-generator pairs produced by rectangular elements oriented at an angle of attack to the flow. The most persistent vortices were produced with an angle of attack of 20 to 30 deg. Edwards and Alker [1974] measured local heat transfer coefficients downstream from a row of cubes, and a row each of corotating and counterrotating vortex generators. The cubes produced the highest local increases in heat transfer; the counterrotating vortex pairs were more effective than the corotating vortices, with a maximum increase in heat transfer of 60% over flat-plate values. Mantle [1966] and Zhang et al. [1989] studied vortex generators in the form of delta wings and cu-

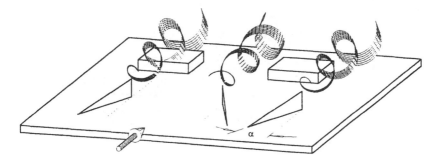

FIGURE 5. Schematic of the single and counter-rotating pair of longitudinal vortices produced by half-delta-wing vortex generators.

bical and cylindrical obstructions. Pearcey [1961] provided design criteria for optimum arrays of vortices including corotating and counterrotating pairs.

Garimella and Eibeck [1991a] reported a study in which half-delta wings were placed upstream from, and on the same wall as, each streamwise column of chips; heat transfer enhancement with water as coolant was studied as a function of streamwise position and Reynolds number. Vortex generators of two heights (one and two times the chip height) were studied. As with other kinds of barriers, the greatest enhancement was observed at the second row of chips downstream and at a Reynolds number in the transitional flow regime, as shown in Figure 6. In a succeeding study by Garimella and Schlitz [1993], half-delta wings were placed on the wall opposite a chip, singly and in counterrotating pairs, and localized heat transfer enhancement on that chip was investigated with the height of the delta wing and its position with respect to the chip as variables. A dielectric liquid (FC-77, with a Prandtl number of 25.3) was used as the coolant. A single, tall vortex generator, twice the chip in height, yielded the best results. A peak enhancement of 17% was obtained at transitional Reynolds numbers. Optimum placement locations for the vortex generators were also identified.

Enhancement techniques may be deduced from the study of the effects of missing elements, height differences between chips on a board, and implanted barriers, on heat transfer and pressure drop in arrays of chips by Sparrow et al. [1982, 1983] and Sparrow and Otis [1985]. The mass (heat) transfer coefficient just downstream from a missing element was found to be 40 to 50% higher as

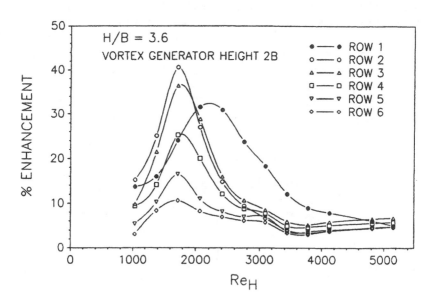

FIGURE 6. Heat transfer enhancement with vortex generators twice the chip in height placed upstream from Row 1, H/B = 3.6 [Garimella and Eibeck, 1991a].

a result. The second row downstream from a barrier implanted in a chip array was found to experience the greatest enhancement (about a factor of 2); however, the pressure drop increased due to the barrier by 10 to 150 times. The location of the second row downstream coincides with the location where the flow reattaches. This is a recurring result that has been found in studies with a variety of obstacles and barriers. The parametric effects of installing multiple barriers and the enhancement on the wall opposite from the chip-carrying wall of a parallel-plates channel have also been studied. The effect of tall elements on heat transfer from the remaining shorter elements in an array was investigated by Souza Mendes and Santos [1987] and Torikoshi et al. [1988]. Chou and Lee [1987] attempted to reduce temperature nonuniformities in air flow over a heated chip using a rectangular barrier mounted at the front face of the downstream chip. The maximum temperature of the chip was reduced as a result, as were the temperature variations over the chip surface. The barrier was found to have an optimum effectiveness for a ratio of barrier to chip height of 2/3, beyond which no increase was found.

Anderson and Moffat [1991] suggested the introduction of scoops (Figure 7a) in the low-velocity recirculation region downstream of each chip to enhance thermal mixing and thus reduce overall temperature rise. A decrease in the component temperature rise of up to 19% was observed as a result of the scoops in the row just downstream (see Figure 7b); the associated increase in pressure drop over the eight streamwise rows of chips with scoops introduced behind one row was found to be 11%. Similarly, Matsushima and Yanagida [1992] measured a 25% increase in heat transfer coefficient in the last of five rows of chips due to cylindrical obstacles placed on either side of each chip in a row, that is, in the flow paths between chips in each spanwise row. With this increase, the heat transfer coefficient was nearly the same as that predicted under the assumption of complete thermal mixing of the cooling air. This arrangement with a total of 10 cylinders placed in the five rows, caused the pressure drop to increase by 1.6 times.

Heat transfer effects due to horseshoe vortices (see Figure 8) along the wall downstream from a wall-mounted cylinder and streamlined cylinder were studied by Fisher and Eibeck [1990], who observed local increases in heat transfer of 20 to 30%. The vortices from a streamlined cylinder were found to be more concentrated (with greater heat transfer augmentation) while those from a circular cylinder were not as strong but were spread over a wider area due to a large region of separation. Local heat and mass transfer effects of horseshoe vortices produced at a cylinder/wall junction were studied by Ireland and Jones [1986] in channel flow and by Goldstein et al. [1985] in an external boundary layer. Ireland and Jones found that the maximum heat transfer coefficient along the channel wall occurs at the stagnation point upstream from the cylinder, and a double peak in Nusselt number occurs downstream. The mass transfer experiments of Goldstein et al. showed strongly enhanced mass transfer immediately upstream (4.5 times the undisturbed value at the leading edge) and along

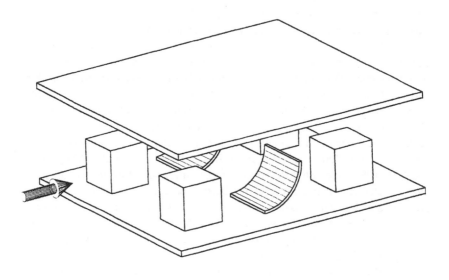

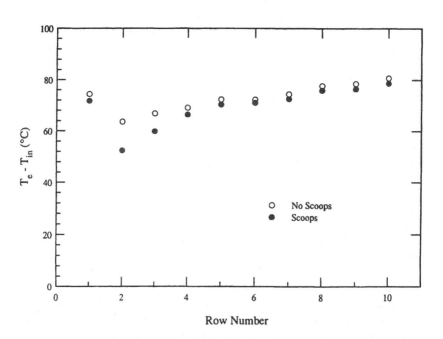

FIGURE 7. (a) Schematic of the scoops installed between two rows of chips; (b) Comparison of the component temperature rise with and without scoops; the scoops are placed between rows 1 and 2 [Anderson and Moffat, 1991].

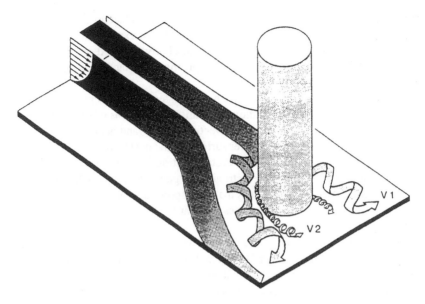

FIGURE 8. Simplified schematic of the horseshoe vortex formed at a wall-mounted cylinder (Goldstein and Karni [1984], by permission).

the sides of the cylinder. In their experiments, a short cylinder with a height to diameter aspect ratio of 1 yielded average mass transfer rates about 8% higher than a taller cylinder (aspect ratio 12) but with markedly different local distributions.

DISPLACED PROMOTERS AND FLOW MODULATION

One technique investigated in the literature to effect heat transfer enhancement has been to mount devices above the chips, either suspended or as an attachment on the opposing wall. These devices serve to locally increase the flow velocity, shed vortices, shift reattachment points, and/or increase turbulent mixing. For instance, Karniadakis et al. [1988] and Thomas [1965, 1966] showed that circular wires positioned above a smooth surface in laminar flow result in a premature transition to turbulence, causing increased heat transfer rates. Myrum et al. [1993] investigated the heat transfer enhancement obtained by placing cylindrical rods immediately above heated ribs in a duct. They defined enhancement in terms of the increase of the entropy generation per unit duct length, and found increases of up to 27% in this quantity due to the introduction of the cylindrical rods. Chou and Chiang [1990] also tested the enhancement due to unsteady vortex shedding from cylindrical rods held above chips, with the flow Reynolds number and the rod location as parameters. A temperature reduction of about 15% was measured at the chip immediately downstream from the chip above which the rod was installed for Reynolds numbers at the high end of the laminar range. At low Reynolds numbers (≤ 100 based on rod diameter), however, the rods had no discernible enhancing effect. Gan et al. [1990] reported results of experiments performed with a half-cylinder and half of a streamlined body mounted on the wall opposite the chips. With two such displaced promoters in tandem, they obtained a peak enhancement of 51%.

Another class of forced convective heat transfer enhancement techniques is based on using flow destabilization mechanisms in regions of separated flow. They promote lateral, large-scale convective mixing and hence increase heat transport normal to the heat transfer surface. Flow unsteadiness is induced by active, passive, or supercritical flow destabilization. It appears that above a critical Reynolds number, flows in electronic chip geometries exhibit laminar, self-sustained oscillations at a frequency that corresponds to that of the most unstable Tollmien-Schlichting channel modes compatible with the periodicity of the geometry.

Active destabilization techniques have been proposed by Sobey [1980], Ghaddar et al. [1986], Greiner [1991], and Azar [1992]. In this approach, the external flow is actively modulated at an appropriate frequency to excite flow instabilities, and thus to disrupt the shear layer and increase communication between stagnant flow regions between chips in an array and the bypass channel flow. Flow destabilization may also be accomplished in a passive mode by using obstacles such as cylinders (displaced promoters) that periodically shed vortices and again serve to disrupt the confined nature of the interchip flow and hence to improve mixing [Karniadakis et al., 1988; Suzuki et al., 1991]. Ratts et al. [1987] studied the flow modulation induced by vortex shedding from cylinders in cross flow; enhancements of up to 82% in heat transfer coefficient were obtained with cylinders placed periodically above the back edge of each row of chips. The cylinder position, diameter, length, and number of cylinders were investigated as parameters.

Active, passive, and supercritical flow destabilization techniques were compared on an equal pumping-power basis by Amon [1992] via numerical simulations. While the passive techniques were found to be the best with respect to minimum power dissipation at low Nusselt numbers, supercritical destabilization was found to become competitive as higher Nusselt numbers became necessary. These results are shown in Figure 9.

Another approach involving the modification of the incoming flow has been to incline the inlet of air in order to intensify the cooling experienced. Inclined inflow could be naturally occurring, for instance, as in the cooling of electric motors where the air enters obliquely into the stators. This approach was explored by Jicha and Horsky [1990] who investigated the effect of the variation of the angle of attack from 0 (axial entrance) to 70 deg for a Reynolds number of 42,520. For longer channels with a length of the order of 15 channel diameters, there was a modest increase in heat transfer of about 15% due to the inclined inlet of air. For short channels (about five diameters long), the increase was a much more dramatic 50% (angle of attack 70 deg).

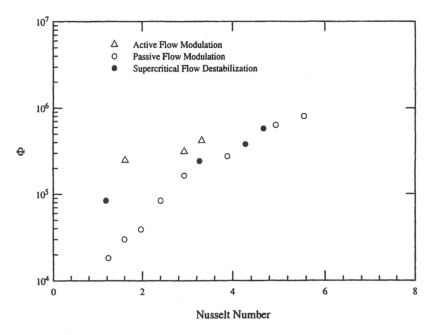

FIGURE 9. Dimensionless pumping power as a function of Nusselt number for various flow modulation techniques [Amon, 1992].

IMPINGING JETS

Impinging jets may be classified as free-surface or submerged, and the spent flow can be in a confined or unconfined state. Also, the nozzles producing the jets could be round orifices or slots, singly or in arrays, based on the application (Figure 10). Free-surface jets are those where the fluid issuing from the jet is different from the ambient fluid, resulting in a distinct free surface separating the two fluids; an example is a jet of water issuing into surrounding air. In submerged jet impingement, on the other hand, the ambient fluid is the same as the impinging jet; this is the case for air-in-air or water-in-water jets.

Four main variables can be identified that affect the local heat transfer from a chip to an impinging jet: jet Reynolds number, fluid thermophysical properties (Prandtl number), nozzle-to-chip spacing, and distance from the stagnation point. A host of other parameters may also be identified including nozzle geometry, thermal and hydrodynamic boundary conditions, turbulence level in the jet, and number and configuration of nozzles. The stagnation-point heat transfer coefficient has been observed to remain constant or to increase slightly as the nozzle-to-chip spacing is increased to about four nozzle diameters and then to decrease with further increases in the spacing. This behavior can be explained based on the fluid mechanics of the jet. As the jet issues from the nozzle, a mixing layer forms over its periphery and penetrates towards the center of the jet with distance traversed away from the nozzle. The so-called potential core of

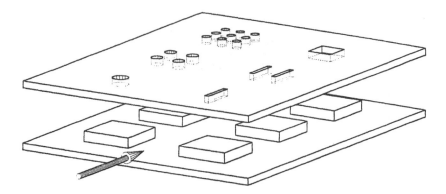

FIGURE 10. Various configurations of round and slot jets, singly and in arrays.

a jet is the region that is unaffected by the mixing layer, and the fluid therein remains at the nozzle exit velocity. For well-formed nozzles, the potential core has been shown to extend over 6 to 8 nozzle diameters, while for square-edged nozzles, the core persists up to only 2 or 3 diameters away. When the spacing between the nozzle and chip is less than the length of the potential core, the core strikes the impingement surface; as long as this condition is met, the stagnation point heat transfer remains roughly constant. For larger spacings, the axial velocity of the jet decreases with increasing distance, and the stagnation heat transfer progressively decreases. However, due to the large-scale structures in the mixing layer of the jet, the level of turbulence in the jet simultaneously increases and compensates for some of the drop in heat transfer due to the drop off in axial velocity.

Much of the experimentation in jet impingement heat transfer has involved air jets, and by default, in the submerged mode, since the experimentally easiest configuration is that of an air jet issuing into air. Air cooling of heated surfaces with impinging jets has been studied extensively [Baughn et al., 1991; Gardon and Akfirat, 1965; Goldstein et al., 1986; Hollworth and Gero, 1985]. Reviews of the literature including design correlations for the use of impinging jets are available in Martin [1977], Obot et al. [1980], and Downs and James [1987].

While a large number of air-jet studies have been reported in the literature, it is important in electronics applications to consider the effect of the confining wall on the heat transfer from a surface. The nozzles used in cooling computer chips would be located in the board facing the chips, and thus the spent flow from the jet would be confined to a parallel-plates channel. The recirculating flows created by the confining wall (in an axisymmetric toroidal shape around the jet centerline for a round jet) have a significant effect on the heat transfer distribution on the target surface as shown by Garimella and Rice [1995] and Rice and Garimella [1994]. Thus care should be employed in extrapolating results from unconfined jets to electronics applications. Figure 11 shows a photograph of the flow patterns in an axisym-

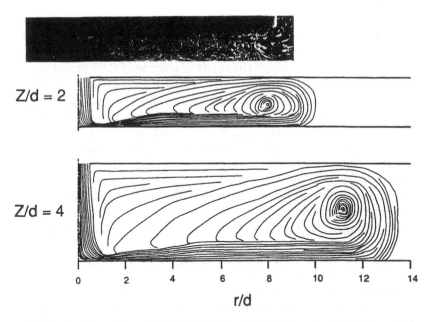

FIGURE 11. Photograph and sketches of the flow patterns in a confined and submerged impinging jet of FC-77 for a nozzle diameter of 3.18 mm and a nozzle to target spacing of 2 (sketch only) and 4 diameters; the photograph and sketches are not to the same scale.

metric, confined, and submerged jet of FC-77 visualized with pliolite particles and laser illumination; also included (with a different scale) are sketches of the flow patterns visualized.

For round jets, the local heat transfer coefficient on the target surface has a bell-shaped distribution with respect to radial distance from the stagnation point. The maximum value occurs at the stagnation point and decreases symmetrically with radial distance. Under some conditions, secondary maxima are observed in these curves corresponding to the location where the wall jet becomes turbulent [den Ouden and Hoogendoorn, 1974; Martin, 1977] and where the recirculating flow reattaches [Garimella and Rice, 1994]. The local heat transfer distribution for a confined and submerged round jet of FC-77 illustrating the secondary peaks for different nozzle to target spacings is shown in Figure 12; similar results have been obtained in air (for instance, Huber and Viskanta [1994]).

One of the implications of using multiple jets to cool a single chip, or single jets to cool an array of chips on a board, especially in a confined configuration, is the resulting cross flow. As the spent fluid from each jet drains away, the accumulated drainage adds a cross-flow component to the existing impinging-jet flow field. The effect of this accumulated cross flow from neighboring jets (or with cross flow imposed) has been studied by Florschuetz et al. [1981], Goldstein and Behbahani [1982], Obot and Trabold [1987], and Whidden et al. [1992]. According to Obot and Trabold, the flow field for these

multiple jet configurations is complicated by interference between neighboring jets prior to impingement, as well as by collision of the developing
wall flows from adjacent nozzles (cross flow). Jiji and Dagan [1987] performed experiments with single (free) jets of water and FC-77 impinging on
a single heat source and an array, with the heat sources oriented vertically.
The effect of cross flow due to the drainage of spent flow from neighboring
jets had only a minor effect on average surface temperature and its unifor-
mity. As in other studies of this type, the recommendation was to increase the
number of jets and decrease the jet diameter for improved heat transfer performance. The interactions between neighboring jets for confined and submerged air jets was recently studied by Huber and Viskanta [1994]. A
nozzle-to-target spacing of 0.25 diameters was shown to result in minimal
preimpingement interaction, thus leading to much improved heat transfer relative to a larger spacing of 6 diameters. The secondary maxima were also
much more pronounced at the smaller spacings (as also shown in Figure 12).
Lytle and Webb [1994] obtained similar results for low nozzle-to-target spacings but in an unconfined configuration.

The literature on jet impingement heat transfer has utilized various fluids
ranging from water to freons in addition to air. It would be advantageous to cast
all these results with a unified perspective so that they could be used irrespective of the fluid used in an application. However, the scaling of results between

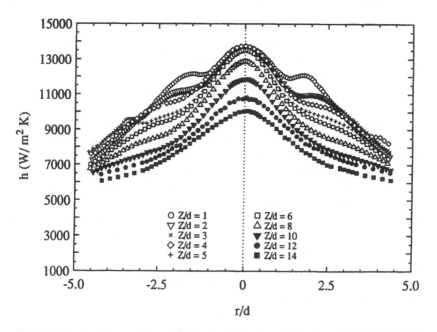

FIGURE 12. Local heat transfer coefficient distribution for a confined, round jet of FC-77 with
a diameter of 1.59 mm, at different nozzle to target spacings and a Reynolds number of 13,000
[Garimella and Rice, 1994].

fluids with different thermophysical properties is not yet well understood, and most studies have simply used assumed Prandtl-number dependencies in their correlations. One study by Metzger et al. [1974] performed experiments with water and oil jets (free-surface) to determine the effect of Prandtl number on heat transfer; a Prandtl number range between 3 and 150 was tested by varying the fluid used and its inlet temperature.

The sizeable literature on the enhancement of jet impingement heat transfer by employing surface roughness, protuberances and fins on the target surface is beyond the scope of this chapter.

COMPACT HEAT SINKS

One of the most compact and high-performance heat dissipation devices is the microchannel heat sink, first proposed by Tuckerman and Pease [1981, 1982] as a technique to lower the convection resistance between the substrate and coolant. The microscopic flow channels incorporated in this heat sink are based on the fact that for laminar flow in confined channels, the convection coefficient scales inversely with channel width. A prototype water-cooled integral heat sink with 50 μm wide and 300 μm deep channels machined into the silicon substrate was shown to maintain a maximum substrate-to-coolant temperature difference of 71°C for a power density of 790 W/cm^2. Clogging of the channels due to entrained debris and the possibility of water leakage were two of the problems that were identified with the implementation of such devices. To address the former, Tuckerman and Pease [1982] suggested that the longitudinal "microfin" structure of the microchannels could be replaced by "micropillars" that would be less prone to clogging by particulate contamination. It was also demonstrated that using FC-77 instead of water as the coolant would lead to only minor degradation of the heat transfer. Phillips et al. [1987] proposed models for evaluating the thermal and fluid performance of such heat sinks. Their numerical predictions showed that turbulent-flow designs could have equivalent or better performance than comparable laminar flow designs. Recently, Bowers and Mudawar [1993] suggested that based upon practical considerations such as pressure drop, erosion, choking, clogging, and manufacturing ease, a minichannel geometry (with channel widths of a few mm) offered inherent advantages over the microchannel geometry, except when the heat dissipation required is beyond the capability of the minichannels, and when minimizing weight and liquid inventory is a must.

Another example of compact heat sinks is the heat pipe. The advantages of the high heat transfer rates obtained through phase-change heat transfer can be coupled with forced convection using a heat pipe, which dissipates heat by a cyclic process involving a coolant that undergoes evaporation and condensation in a contained unit (for instance, see Dunn and Reay [1994]). In a heat pipe, the inherent problems associated with direct liquid cooling and boiling at the chip surface are avoided while high heat fluxes are still achieved. The advantage of using a heat pipe lies in its ability to carry heat from densely populated areas through a very narrow *conduit* to a location where a larger heat sink can be installed. Thus a fin array can be used at a location remote from the chip, which would otherwise have been too bulky to install. Heat pipes may also be used as heat spreaders for controlling temperature nonuniformities on the surface of semiconductor devices, as in the "micro" heat pipes of Wu et al. [1990].

At the normal operating temperatures of chips, typically less than 100°C, Babin and Peterson [1990] achieved heat dissipation rates of over 125 W/cm² with a maximum total power of 65 W. Unlike other enhancement techniques where the heat flux can be used to calculate the power dissipated, the maximum power that can be carried away by a heat pipe is governed by the capabilities of the phase change process. A heat-pipe design for cooling high flux/high power chips was recently explored by North and Avedisian [1993]. The design involved a series of holes drilled into a manifold base plate lined with sintered copper powder that served as the wick. With an air-cooled condenser section, a maximum heat flux of 47 W/cm² and a total power of 900 W were achieved with surface temperatures under 100°C.

While a heat pipe represents a very high thermal-conductivity path between the chip and the ultimate heat sink such as a cold plate or a fin array, it may often be limited by the capability of this heat sink to dissipate the heat that is wicked to it. Thus, while large amounts of heat can be carried through the heat pipe, impractically large condenser surface areas may be necessary to manage this heat at the sink. The main impediments to the widespread use of heat pipes has been their high manufacturing cost and uncertainties about their long-term reliability. With stepped-up research and development efforts, improved manufacturing methods, and increased heat-dissipation needs they are fast becoming a viable design choice.

POROUS METALLIC MATRICES

Porous metallic inserts have been proposed for service as high-efficiency fins. The microchannel cooling schemes discussed above could be considered an extreme application of this technique. The microstructures have been modeled as a porous medium with Darcy's law being used to describe the flow [Koh and Colony, 1986]. However, as pointed out in Vafai and Tien [1981], heat convection problems may not accurately be modeled using Darcy's law. Using a modified Darcy equation for the flow and volume-averaged energy equations for the solid and fluid, Tien and Kuo [1987] proposed a model that could be applied not only to microchannels and dense porous inserts but also to other complex nonhomogeneous systems such as pin-fin structures. A foam-filled duct was shown to yield two to four times the heat transfer relative to laminar slug flow in a parallel-plates duct. Hadim and Bethancourt [1993] showed that, instead of completely filling the duct with porous material, discrete porous inserts mounted only above each heat source resulted in significantly lower pressure drops, while the heat transfer enhancement effects did not degrade noticeably.

IMPROVED SUBSTRATE CONDUCTION

One approach, which does not fall into the traditional classification of enhancement techniques but is of great interest due to its passivity, is improved substrate conduction. Since the choice of board materials, die-attach and bonding techniques are not primarily governed by thermal management considerations, this approach may not always be practicable. However, if thermal concerns were to be addressed at an early design stage, the improvement of substrate conduction could be considered an enhancement parameter. Antonetti [1990] reviewed the progress made in the development of predictive theory, recent experimental studies, and examples of thermal interface management in current electronic equipment. The theory and the applications of constriction and spreading resistance concepts as applied to microelectronic thermal management were reviewed by Yovanovich [1987]. Through their models Culham and Yovanovich [1987] and Negus and Yovanovich [1986] illustrated that the thermal conductivity is the single most important design parameter for reducing board temperatures. A marginal increase in the relative percentage of copper in the board was shown to result in an increased effective thermal conductivity (for board conductivity in the range of 1 to 10 W/mK) and significantly lower board temperatures. The effect on board temperature of the small alterations in circuit-board thickness was shown to be not sufficient to warrant changing the design of the board. It was also demonstrated that a decrease in thermal resistance of the board due to increased chip spacing occurs only when accompanied by a similar increase in the board thickness (which is often not feasible). Thus increasing the chip spacing to a distance greater than twice the chip width would produce little benefit in terms of decreased thermal resistance.

HYBRID TECHNIQUES

Some of the enhancement techniques discussed in this chapter may be used in combination with each other. Air cooling is often supplemented by liquid cooling as the ultimate heat sink. For instance, water-cooled finned-tube heat exchangers are incorporated between successive rows of circuit boards. In the IBM Liquid Encapsulated Module (LEM), the substrate carrying the integrated circuit chips is mounted within a sealed module-cooling assembly containing a fluorocarbon liquid coolant. Heat from the chips is transferred to the fluorocarbon using internal fins and then to water flowing through an externally attached cold plate. One of the more successful designs for indirect liquid cooling was the Thermal Conduction Module (TCM), developed at IBM and used extensively in their computers. These and other such techniques are discussed in Chu and Simons [1990].

PERFORMANCE EVALUATION CRITERIA

In a given application, the performance evaluation criteria may include increase in heat duty, increase in surface area, pumping power, initial cost, maintenance cost, safety, and reliability. These criteria are discussed in Bergles et al. [1974] and can be variously applied to optimize heat transfer based on the particular application in question. For instance, Bejan and Morega [1994] proposed methods of calculating the optimal spacing of circuit boards in turbulent forced convection for a stack of fixed volume by maximizing the overall thermal conductance. Mikic et al. [1990] addressed the optimization problem with respect to the selection of the channel hydraulic diameter and flow Reynolds number to yield the minimum pumping power for a given convective cooling configuration. Although this is a very important part of any study of enhancement techniques, it is beyond the scope of this chapter; the reader is referred to the review articles on enhancement techniques cited earlier as well as Chapter 1 in this book for further information.

CONCLUSION

All of the techniques discussed in this chapter have focussed on enhancing air cooling. The focus has been on those methods that are particularly attractive in the electronics-packaging field, and thus, techniques such as the application of vibration (either in the fluid or at the substrate) or of electric fields have not been included. Also, while air cooling is mechanically the simplest alternative, the continuing advances in miniaturization and packaging have already necessitated non-air-based cooling technologies in several applications; such technologies are likely to see more widespread use in the near future.

ACKNOWLEDGMENTS

The author's early work in this area at the University of California at Berkeley (with P. A. Eibeck) was supported by IBM Corporation and the National Science Foundation; ongoing work at the University of Wisconsin-Milwaukee is supported primarily by Cray Research, Inc. This support is gratefully acknowledged.

APPENDIX A: NOMENCLATURE

A   exposed (active) surface area of chip
B   chip height
d, D   nozzle diameter
h   heat transfer coefficient
H   channel height
Mo   Mouromtseff number
Nu   Nusselt number
Re_H Reynolds number based on channel height
Re_L Reynolds number based on chip length
T_e element (chip) temperature

T_{in} channel mean-inlet temperature
U   channel inlet velocity
Z   nozzle to target spacing
ϕ dimensionless pumping power $\left(\phi = \dfrac{3}{4} \dfrac{dp}{dx} \dfrac{Vh^4}{\rho v^3} \right)$
v kinematic viscosity

REFERENCES

Aihara, T. 1991. Air cooling techniques by natural convection, in *Cooling Techniques for Computers*, W. Aung, Ed., Hemisphere, New York, 1–45.

Amon, C.H. 1992. Heat transfer enhancement by flow destabilization in electronic chip configurations, *J. Electron. Packag.*, 114:35–40.

Anderson, A.M. and Moffat, R.J. 1991. Direct air cooling of electronic components: reducing component temperatures by controlled thermal mixing, *J. Heat Transfer*, 113:56–62.

Antonetti, V.W. 1990. Thermal Contact Resistance in Electronic Equipment, in *Heat Transfer in Electronic and Microelectronic Equipment*, A.E. Bergles, Ed., Hemisphere, Washington D.C., 663–673.

Ashiwake, N., Nakayama, W., Daikoku, T., and Kobayashi, F. 1983. Forced convective heat transfer from LSI packages in an air-cooled wiring card array, *Heat Transfer in Electronic Equipment*, ASME HTD, 28:35–42.

Azar, K. 1992. Enhanced cooling of electronic components by flow oscillation, *J. Thermophys. and Heat Transfer*, 6(4):700–706.

Azar, K., Develle, S.E., and Manno, V.P. 1989. Sensitivity of circuit pack thermal performance to convective and geometric variation, *IEEE Trans. Components, Hybrids, and Mfg. Technol.*, 12:732–740.

Azar, K. and Russell, E.T. 1991. Effect of component layout and geometry on the flow distribution in electronics circuit packs, *J. Electron. Packag.*, 113:50–57.

Babin, B.R. and Peterson, G.P. 1990. Experimental investigation of a flexible bellows heat pipe for cooling discrete heat sources, *J. Heat Transfer*, 112:602–607.

Baughn, J.W., Hechanova, A.E., and Yan, X. 1991. An experimental study of entrainment effects on the heat transfer from a flat surface to a heated circular impinging jet, *J. Heat Transfer*, 3:143–149.

Baumann, W. and Rehme, K. 1975. Friction correlations for rectangular roughness, *Int. J. Heat Mass Trans.*, 18:1189–1197.

Bejan, A. and Morega, A.M. 1994. The optimal spacing of a stack of plates cooled by turbulent forced convection, *Int. J. Heat Mass Transfer*, 37:1045–1048.

Bergles, A.E., Blumenkrantz, A.R., and Taborek, J. 1974. Performance evaluation criteria for enhanced heat transfer surfaces, *Japan Soc. Mech. Eng.*, 2:239–243.

Bergles, A.E., Jensen, M.K., Somerscales, E.F.C., and Manglik, R.M. 1991. Literature review of heat transfer enhancement technology for heat exchangers in gas fired applications, Report GRI 91-0146, Gas Research Institute, Chicago, IL.

Bergles, A.E., Nirmalan, V., Junkhan, G.H., and Webb, R.L. 1983. Bibliography on augmentation of convective heat and mass transfer II. Heat Transfer Laboratory Report HTL-31, ISU-ERI-Ames-84221, Iowa State University, December.

Bhatti, M.S. and Shah, R.K. 1987. Turbulent and transition flow convective heat transfer in ducts, in *Single Phase Convective Heat Transfer Handbook*, Wiley-Interscience, New York.

Blair, M.F. 1983a. Influence of free-stream turbulence on turbulent boundary layer heat transfer and mean profile development. I. Experimental data, *J. Heat Transfer*, 105:33–40.

Blair, M.F. 1983b. Influence of free-stream turbulence on turbulent boundary layer heat transfer and mean profile development. II. Analysis of results, *J. Heat Transfer*, 105:41.

Bowers, M.B. and Mudawar, I. 1993. Two-phase electronic cooling using mini-channel and microchannel heat sinks. II. Flow rate and pressure drop constraints, *Adv. in Electron. Packag.*, ASME EEP. 4, 2:703–712.

Castro, I.P. and Robins, A.G. 1977. The flow around a surface-mounted cube in uniform and turbulent streams, *J. Fluid Mech.*, 79:307–335.

Chou, J.H. and Chiang, K.F. 1990. Thermal enhancement for simulated LSI packages by external rods, in *Heat Transfer in Electronic and Microelectronic Equipment*, A. E. Bergles, ed., Hemisphere, Washington, D.C., 393–404.

Chou, J.H. and Lee, J. 1987. Reducing flow non-uniformities in LSI packages by vortex generators, in *Proc. Int. Symp. Cooling Technology for Electron. Equipment*, Hawaii, 583–594.

Chu, R.C., Seely, J.H., Antonetti, V.W., and Pascuzzo, A.L. 1970. Thermal design optimization in large digital systems, IBM Technical Report, TR 00.2039, June.

Chu, R.C. and Simons, R.E. 1990. Evolution of cooling technology in medium and large scale computers—an IBM perspective, in *Heat Transfer in Electronic and Microelectronic Equipment*, A. E. Bergles, ed., Hemisphere, Washington D.C., 41–60.

Crabb, D., Durao, D.F.G., and Whitelaw, J.H. 1977. Velocity characteristics in the vicinity of a two-dimensional rib, in *Proc. 4th Brazilian Congress of Mech. Eng.*, Paper B-3, 415–429.

Culham, J.R. and Yovanovich, M.M. 1987. Non-iterative technique for computing temperature distributions in flat plates with distributed heat sources and convective cooling, in *Proc. ASME/JSME Heat Transfer Conf.*, Hawaii, 403–409, March.

Dalle Donne, M. and Meyer, L. 1977. Turbulent convective heat transfer from rough surfaces with two-dimensional rectangular ribs, *Int. J. Heat Mass Transfer*, 20:583–620.

den Ouden, C. and Hoogendoorn, C.J. 1974. Local convective heat transfer coefficient for jets impinging on a plate: Experiments using a liquid crystal technique, in *Proc. 5th Int. Heat Transfer Conf.*, Tokyo, Japan, September 3–7.

Dipprey, D.F. and Sabersky, R.H. 1963. Heat and momentum transfer in smooth and rough tubes at various Prandtl numbers, *Int. J. Heat Mass Transfer*, 6:329–353.

Downs, S.J. and James, E.H. 1987. Jet impingement heat transfer—a literature survey, ASME Paper 87-HT-35.

Dunn, P.D. and Reay, D.A. 1994. *Heat Pipes*, 4th ed., Pergamon Press, New York.

Durst, F., Founti, M., and Obi, S. 1986. Experimental and computational investigation of the two-dimensional channel flow over two fences in tandem, in *3rd Int. Symp. on Applications of Laser Anemometry to Fluid Mechanics*, Paper 2.5, Lisbon, July.

Eaton, J.K. and Johnston, J.P. 1981. A review of research on subsonic turbulent flow reattachment, *AIAA J*, 19:1093–1100.

Edwards, F.J. and Alker, C.J.R. 1974. The improvement of forced convection surface heat transfer using surface protrusions in the form of (A) cubes and (B) vortex generators, in *Proc. 5th Int. Heat Transfer Conf.*, Tokyo, 244–248.

Eibeck, P.A. and Eaton, J.K. 1987. Heat transfer effects of a longitudinal vortex embedded in a turbulent boundary layer. *J. Heat Transfer*, 109:16–24.

Fiebig, M., Kallweit, P., and Mitra, N.K. 1986. Wing type vortex generators for heat transfer enhancement, in *Proc. 8th Int. Heat Transfer Conf.*, San Francisco, 6:2909–2913.

Fisher, E.M. and Eibeck, P.A. 1990. The influence of a horseshoe vortex on local convective heat transfer, *J. Heat Transfer*, 112:329–335.

Florschuetz, L.W., Truman, C.R., Metzger, D.E. 1981. Streamwise flow and heat transfer distributions for jet array impingement with crossflow, *J. Heat Transfer*, 103:337–342.

Gan, Y.P., Wang, S. Lei, D.H. and Ma, C.F. 1990. Enhancement of forced convection air cooling of block-like electronic components in in-line arrays, in *Heat Transfer in Electronic and Microelectronic Equipment*, A. E. Bergles, Ed., Hemisphere, Washington, D.C., 223–233.

Gardon, R. and Akfirat, J.C. 1965. The role of turbulence in determining the heat transfer characteristics of impinging jets, *Int. J. Heat Mass Transfer*, 8:1261–1272.

Garimella, S.V. 1991. Physical mechanisms for the local heat transfer enhancement caused by fin-like obstacles in heat exchanger flow passages, *SAE 1991 Transactions, J. of Mater. and Mfg.*, 100(5):208–217.

Garimella, S.V. and Eibeck, P.A. 1990. Heat transfer characteristics of an array of protruding elements in single phase forced convection, *Int. J. Heat Mass Transfer*, 33(12):2659–2669.

Garimella, S.V. and Eibeck, P.A. 1991a. Enhancement of single phase convective heat transfer from protruding elements using vortex generators, *Int. J. Heat Mass Transfer*, 34(9):2427–2430.

Garimella, S.V. and Eibeck, P.A. 1991b. Effect of spanwise spacing on the heat transfer from an array of protruding elements in forced convection, *Int. J. Heat Mass Transfer*, 34(9):2431–2433.

Garimella, S.V. and Eibeck, P.A. 1991c. Fluid dynamic characteristics of the flow over an array of large roughness elements, *J. Electron. Packag.*, 113(4):367–373.

Garimella, S.V. and Eibeck, P.A. 1992. Onset of transition in the flow over a three-dimensional array of rectangular obstacles, *J. Electron. Packag.*, 114(2):251–255.

Garimella, S.V. and Rice, R.A. 1995. Confined and submerged liquid jet impingement heat transfer, *J. Heat Transfer*, 117(4): 871–877.

Garimella, S.V. and Schlitz, D.J. 1993. Reducing inter-chip temperature differences in computers using vortex generators in forced convection, *J. Electron. Packag.*, 115(4):410–415.

Garimella, S.V. and Schlitz, D.J. 1995. Heat transfer enhancement in narrow channels using two and three-dimensional mixing devices, *J. Heat Transfer*, 117(3): 590–596.

Garratt, J.R. and Hicks, B.B. 1973. Momentum, heat, and water vapor transfer to and from natural and artificial surfaces, *Quar. J. Roy. Met. Soc.*, 99:680–687.

Ghaddar, N.K., Korczak, K.Z., Mikic, B.B., and Patera, A.T. 1986. Numerical investigation of incompressible flow in grooved channels. II. Resonance and oscillatory heat-transfer enhancement, *J. Fluid Mech.*, 168:541–567.

Goldstein, R.J. and Behbahani, A.I. 1982. Impingement of a circular jet with and without cross flow, *Int. J. Heat Mass Transfer*, 25:1377–1382.

Goldstein, R.J. and Karni, J. 1984. The effect of a wall boundary layer on local mass transfer from a cylinder in crossflow, *J. Heat Transfer*, 106:260–267.

Goldstein, R.J., Chyu, M.K., and Hain, R.C. 1985. Measurement of local mass transfer on a surface in the region of the base of a protruding cylinder with a computer-controlled data acquisition system, *Int. J. Heat Mass Transfer*, 28(5):977–985.

Goldstein, R.J., Behbahani, A.I., and Kieger Heppelmann, K. 1986. Streamwise distribution of the recovery factor and the local heat transfer coefficient to an impinging circular air jet, *Int. J. Heat Mass Transfer*, 29:1227–1235.

Greiner, M. 1991. An experimental investigation of resonant heat transfer enhancement in grooved channels, *Int. J. Heat Mass Transfer*, 34(6):1383–1391.

Hadim, A. and Bethancourt, A. 1993. Numerical study of forced convection in a partially porous channel with localized heat sources, *Adv. in Electron. Packag.*, ASME, EEP-4-2:649–657.

Hall, W.B. 1962. Heat transfer in channels having rough and smooth surfaces, *J. Mech. Eng. Sci.*, 4:287–291.

Han, J.C., Glicksman, L.R., and Rohsenow, W.M. 1978. An investigation of heat transfer and friction characteristics for rib-roughened surfaces, *Int. J. Heat Mass Transfer*, 21:1143–1156.

Hollworth, B.R. and Fuller, H.A. 1987. Heat transfer and pressure drop in a staggered array of air cooled components, in *Proc. Int. Symp. Cooling Technol for Electron Equipment*, Honolulu, Hawaii, 732–748.

Hollworth, B.R. and Gero, L.R. 1985. Entrainment effects on impingement heat transfer: Part II— Local heat transfer measurements, *J. Heat Transfer*, 107:910–915.

Huber, A.M. and Viskanta, R. 1994. Effect of jet-jet spacing on convective heat transfer to confined, impinging arrays of axisymmetric air jets, *Int. J. Heat Mass Transfer*, 37:2859–2869.

Igarashi, T. and Takasaki, H. 1992. Fluid flow around three rectangular blocks in a flat-plate laminar boundary layer, *Experimental Heat Transfer*, 5:17–31.

Ireland, P.T. and Jones, T.V. 1986. Detailed measurements of heat transfer on and around a pedestal in fully developed passage flow, in *Procs. 8th Int. Heat Transfer Conf.*, C. L. Tien et al. Eds., 3:975–980.

Jicha, M. and Horsky, J. 1990. Intensification of air cooling by inclined inlet of air, in *Heat Transfer in Electronic and Microelectronic Equipment*, A. E. Bergles, Ed., Hemisphere, Washington, D.C., 365–380.

Jiji, L.M. and Dagan, Z. 1987. Experimental investigation of single phase multi-jet impingement cooling of an array of microelectronic heat sources, in *Procs. Int. Symp. Cooling Technol. for Electron. Equipment*, Honolulu, Hawaii, 265–283.

Kader, B.A. and Yaglom, A.M. 1977. Turbulent heat and mass transfer from a wall with parallel roughness ridges, *Int. J. Heat Mass Transfer*, 20:345–357.

Karniadakis, G.E., Mikic, B.B., and Patera, A.T. 1988. Minimum dissipation transport enhancement by flow destabilization: Reynolds' analogy revisited, *J. Fluid Mech.*, 192:365–391.

Koh, J.C.Y and Colony, R. 1986. Heat transfer microstructures for integrated circuits, *Int. Comm. Heat Mass Transfer*, 13:89–98.

Kraus, A.D. and Bar-Cohen, A. 1983. *Thermal Analysis and Control of Electronic Equipment*, McGraw-Hill, New York, chap. 19.

Lim, F.K. and Lewkowicz, A.K. 1986. Investigation by laser Doppler anemometry of the turbulent flow around cylindrical obstacles on a rough surface, *Int. J. Heat Fluid Flow*, 7:102–108.

Lytle, D. and Webb, B.W. 1994. Air jet impingement heat transfer at low nozzle-plate spacings, *Int. J. Heat Mass Transfer*, 37:1687–1697.

Maciejewski, P.K. and Moffat, R.J. 1989. Heat transfer with very high free stream turbulence, Report HMT-42, Stanford University, Stanford, CA.

Mantle, P.L. 1966. A new type of roughened heat transfer surface selected by flow visualization techniques, in *Proc. 3rd Int. Heat Transfer Conf.*, 1:45–55.

Martin, H. 1977. Heat and mass transfer between impinging gas jets and solid surfaces, *Adv. Heat Transfer*, 13:1–60.

Matsushima, H. and Yanagida, T. 1992. Heat transfer from finned LSI packages in a channel between circuit boards, *Heat Transfer-Japanese Res.*, 21(2):165–176.

Maubach, K. 1972. Rough annulus pressure drop—interpretation of experiments and recalculation for square ribs, *Int. J. Heat Mass Transfer*, 15:2489–2498.

Metzger, D.E., Cummings, K.N., and Ruby, W.A. 1974. Effects of Prandtl number on heat transfer characteristics of impinging liquid jets, in *Proc. 5th Int. Heat Transfer Conf*, II:20–24.

Meyer, L. 1980. Turbulent flow in a plane channel having one or two rough walls, *Int. J. Heat Mass Transfer*, 23:591–608.

Mikic, B.B., Kozlu, H., and Patera, A.T. 1990. A methodology for optimization of convective cooling systems for electronic devices, in *Heat Transfer in Electronic and Microelectronic Equipment*, A. E. Bergles, Ed., Hemisphere, Washington, D.C., 269–286.

Moffat, R.J. and Ortega, A. 1989. Direct air-cooling of electronic components, in *Advances in Thermal Modeling of Electronic Components and Systems*, Vol. I, A. Bar-Cohen and A. D. Kraus, Eds., Hemisphere, New York, chap. 3.

Morris, G.K., Garimella, S.V., and Hlavac, D.L. 1995. Measurements of the wake behind a heated obstacle in forced convection, to be presented at the National Heat Transfer Conference, Portland, Oregon.

Myrum, T.A., Qiu, X., and Acharya, S. 1993. Heat transfer enhancement in a ribbed duct using vortex generators, *Int. J. Heat Mass Transfer*, 36:3497–3508.

Nakayama, W. 1982. Enhancement of heat transfer, heat transfer-1982, in *Proc. 7th Int. Heat Transfer Conf.*, Munich, Germany, 1:223–240.

Negus, K.J. and Yovanovich, M.M. 1986. Thermal analysis and optimization of convectively cooled microelectronic circuit boards, *Heat Transfer in Electron. Equipment*, ASME HTD, 57:167–176.

Nikuradse, J. 1933. Laws for flow in rough pipes, *NACA TM*, 1292, (English translation, 1950).

North, M.T. and Avedisian, C.T. 1993. Heat pipes for cooling high flux/high power semiconductor chips, *J. Electron. Packag.*, 115:112–117.

Obot, N.T., Mujumdar, A.S., and Douglas, W.J.M. 1980. Design correlations for heat and mass transfer under various turbulent impinging jet configurations, *Drying*, 388–402.

Obot, N.T. and Trabold, T.A. 1987. Impingement heat transfer within arrays of circular jets. Part I—effects of minimum, intermediate, and complete crossflow for small and large spacings, *J. Heat Transfer*, 109:872–879.

Osborne, D.G. and Incropera, F.P. 1985. Experimental study of mixed convection heat transfer for transitional and turbulent flow between horizontal, parallel plates, *Int. J. Heat Mass Transfer*, 28(7):1337–1344.

Pearcey, H.H. 1961. Shock-induced separation and its prevention by design and boundary layer control. IV. *Boundary Layer and Flow Control, Its principles and application*. G. V. Lachmann, Ed., Pergamon Press, New York.

Pedisius, A.A., Kazimekas, P.V.A., and Slanciauskas, A.A. 1979. Heat transfer from a plate to a high-turbulence air flow, *Soviet Res.*, 11(5):125–.

Perry, A.E., Schofield, W.H., and Joubert, P.H. 1969. Rough wall turbulent boundary layers, *J. Fluid Mech.*, 37:193–211.

Phillips, R.J., Glicksman, L.R., and Larson, R. 1987. Forced-convection, liquid-cooled, microchannel heat sinks for high-power-density microelectronics, in *Proc. Int. Symp. Cooling Technol. for Electron. Equipment*, Honolulu, Hawaii, 227–248.

Ratts, E., Amon, C.H., Mikic, B.B., Patera, A.T. 1987. Cooling enhancement of forced convection air cooled chip array through flow modulation induced by vortex-shedding cylinders in crossflow, in *Proc. Int. Symp. Cooling Technol. for Electron. Equipment*, Honolulu, Hawaii, 651–662.

Rice, R.A. and Garimella, S.V. 1994. Heat transfer from discrete heat sources using an axisymmetric, submerged and confined liquid jet, in *Proc. Int. 10th Heat Transfer Conf.*, Brighton, UK 3:89–94.

Russell, C.M.B., Jones, T.V., and Lee, G.H. 1982. Heat transfer enhancement using vortex generators, in *Proc. 7th Int. Heat Transfer Conf.*, Munich, 3:283–288.

Simonich, J.C. and Bradshaw, P. 1978. Effect of free-stream turbulence on heat transfer through a turbulent boundary layer, *J. Heat Transfer*, 100:673.

Sobey, I.J. 1980. On flow through furrowed channels. I. Calculated flow patterns, *J. Fluid Mech.*, 96:1–26.

Souza Mendes, P.R. and Santos, W.F.N. 1987. Heat transfer and pressure drop experiments in air-cooled electronic-component arrays, *J. Thermophys. and Heat Transfer*, 1(4):373–378.

Sparrow, E.M., Niethammer, J.E., and Chaboki, A. 1982. Heat transfer and pressure drop characteristics of arrays of rectangular modules encountered in electronic equipment, *Int. J. Heat Mass Transfer*, 25:961–973.

Sparrow, E.M. and Otis, D.R. 1985. Ductflow heat transfer at a smooth wall which faces a wall covered by protuberances, *Int. J. Heat Mass Transfer*, 28:1317–1326.

Sparrow, E.M., Vemuri, S.B., and Kadle, D.S. 1983. Enhanced and local heat transfer, pressure drop, and flow visualization for arrays of block-like electronic components, *Int. J. Heat Mass Transfer*, 26:689–699.

Suzuki, H., Kida, S., Nakamae, T., and Suzuki, K. 1991. Flow and heat transfer over a backward-facing step with a cylinder mounted near its top corner, *Int. J. Heat Fluid Flow*, 12(4):353–359.

Thomas, D.G. 1965. Forced convection mass transfer. Part II. effect of wires located near the edge of the laminar boundary layer on the rate of forced convection from a flat plate, *AIChE J.*, 11:848–852.

Thomas, D.G. 1966. Forced convection mass transfer. Part III. Increased mass transfer from a flat plate caused by the wake from cylinders located near the edge of the boundary layer, *AIChE J.*, 12:124–130.

Tien, C.L. and Kuo, S.M. 1987. Analysis of forced convection in microstructures for electronic system cooling, in *Proc. Int. Symp. Cooling Technol. for Electron. Equipment*, Honolulu, Hawaii, 217–226.

Torikoshi, K., Kawazoe, M., and Kurihara, T. 1988. Convective heat transfer characteristics or arrays of rectangular blocks affixed to one wall of a channel, *Natural and Mixed Convection in Electronic Equipment Cooling*, ASME HTD, 100:59–65.

Townes, H.W. and Sabersky, R.H. 1966. Experiments on the flow over a rough surface, *Int. J. Heat Mass Transfer*, 9:729–738.

Tropea, C.D. and Gackstatter, R. 1985. The flow over two-dimensional surface-mounted obstacles at low Reynolds numbers, *J. Fluids Eng.*, 107:489–494.

Tuckerman, D.B. and Pease, R.F.W. 1981. High-performance heat sinking for VLSI, *IEEE Electron Device Lett.*, EDL-2(5):126–129.

Tuckerman, D.B. and Pease, R.F. 1982. Ultrahigh thermal conductance microstructures for cooling integrated circuits, in *Proc. 32nd Electron. Components Conf.*, IEEE, EIA, CHMT, 145–149.

Vafai, K. and Tien, C.L. 1981. Boundary and inertia effects on flow and heat transfer in porous media, *Int. J. Heat Mass Transfer*, 24:195–203.

Wassel, A.T. and Mills, A.F. 1979. Calculation of variable property turbulent friction and heat transfer in rough pipes, *J. Heat Transfer*, 101:469–474.

Webb, R.L. 1987. Enhancement of Single-Phase Heat Transfer, in *Single Phase Convective Heat Transfer Handbook*, Wiley-Interscience, New York, chap. 17.

Webb, R.L. 1994. *Principles of Enhanced Heat Transfer*, John Wiley and Sons, New York.

Webb, R.L., Eckert, E.R.G., and Goldstein, R.J. 1971. Heat transfer and friction in tubes with repeated-rib roughness, *Int. J. Heat Mass Transfer*, 14:601–617.

Wenthen, F.T. 1977. Experimental verification, presented at NSF Research Workshop on Directions of Heat Transfer in Electronic Equipment, Atlanta, Georgia, October.

Whidden, G.L., Stevens, J., and Webb, B.W. 1992. Heat transfer and flow characteristics of two-dimensional jets impinging on heated protrusions with crossflow of the spent air, *J. Electron. Packag.*, 114:81–87.

Wroblewski, D.E. and Eibeck, P.A. 1991. Measurements of the turbulent heat transport in a boundary layer with an embedded streamwise vortex, *Int. J. Heat Mass Transfer*, 34:1617–1631.

Wu, D., Peterson, G.P., and Chang, W. 1990. Experimental investigation of the transient behavior of micro heat pipes, AIAA Paper 90, 1791.

Yovanovich, M.M. 1987. Theory and applications of constriction and spreading resistance concepts for microelectronic thermal management, in *Proc. Int. Symp. Cooling Technol. for Electron. Equipment*, Honolulu, Hawaii, 1–80.

Zhang, Fiebig, M., and Mitra, N.K. 1989. Vortex breakdown on heat transfer enhancement in flows between parallel plates, Eurotherm Seminar 91, 97–104, *Heat Transfer in Single Phase Flows*, Proceeding from seminar in Bochum, Germany.

Chapter 6

LIMITS OF AIR COOLING—
A METHODICAL APPROACH

Kaveh Azar

CONTENTS

0-8493-9447-3/96/$0.00+$.50
© 1996 by CRC Press, Inc.

INTRODUCTION

Most electronic systems are cooled by air because of its availability and ease of implementing. Cooling is done either by natural or forced convection depending upon the system's requirements and application. Cooling system designs vary from simple natural convection to high-capacity forced convection. In natural convection cooled systems, the circulation of air is accommodated by placement of vent ports, e.g., a VCR, along with proper conduction heat transfer paths to disperse the heat from the critical areas. In forced convection systems, fan is the most common fluid mover used. In high-capacity forced convection cooled systems, jet impingement along with surface enhancement (heat sink) is used [Bar-Cohen, 1987]. Typically, these systems have a great deal of customization that is the inherent nature of them; thus, they are not as generic as placing fans in the system.

In a typical design process, one confronts the question of which cooling system or method is most suitable. Typically, a design criterion relating to junction temperature or industry standards is considered, e.g., NEBS for telecommunication equipment [NEBS, 1993]. The desire to cool all systems with natural convection continues to persist, but adherence to specifications and limited cooling capacity is a point of contention. Likewise, cooling with forced convection may not be feasible and high-capacity cooling, i.e., liquid cooling, may be considered for a given system. The question is: What are the limits of air cooling and at what point must higher-end cooling be used? In this chapter, I attempt to answer this question and provide mechanisms for predictions. In addition, I will attempt to highlight the important parameters that contribute to the performance of cooling with air.

To achieve the goal of the chapter, I first review the thermal phenomenon in electronics systems. The intent of this section is to highlight the importance of starting thermal management at system level versus individual component or circuit board. It further elevates the need to look at the component junction temperature instead of total heat dissipation of the system or circuit pack. Hence, thermal coupling and transport is a point of interest and needs to be understood. Then the chapter focuses on the concept of limit, and why it is important to look at the junction temperature instead of the heat flux as the limiting factor. The two subsequent sections show a first-order model that shows heat transfer from a circuit board and a component. These models allow us to highlight the important parameters that impact heat transfer from a component and must be considered in the design process. The last two sections deal with the potential limits of cooling with natural and forced convection when air is the cooling fluid.

THERMAL PHENOMENON
IN ELECTRONIC ENCLOSURES

Thermal phenomenon or process defines the procedure for removal of heat from the components. Thermal process is defined as the merger of heat transfer and fluid flow to transport the energy. The very nature of the thermal phenomenon is then a function of the mode of heat transfer and the fluid flow regime in a system. Thus, proper definition of thermal phenomenon requires understanding of transport mechanisms—heat transfer and fluid flow.

In this section, I start by defining the basic principles of heat transfer. Then heat transfer in electronic components is discussed. Here, I discuss how the power, being dissipated in the chips, is eventually transferred to the cooling fluid. Since circuit packs by definition contain several components and play an important role in thermal response, thermal transport in circuit packs (PCB) are then discussed. The last two sections talk about the thermal coupling (communication) between elements that form an electronic enclosure (system).

HEAT TRANSFER MECHANISMS

There are three modes of heat transfer: conduction, convection, and radiation. Conduction heat transfer is when the heat is transferred by molecular vibration—solids or stagnant fluids. An example of a solid is the molding compound or the substrate in a component, and stagnant fluid is the air trapped between the component and circuit board. The conduction heat transfer is governed by Fourier cooling law defined as

$$Q = kA (T_h - T_c)/L \tag{1}$$

In this equation, k is the thermal conductivity and is a property of the material. Table 1 shows the thermal conductivity of typical materials used in components.

Convection heat transfer is when the transport of heat takes place by fluid motion. Three types of convection heat transfer are recognized: natural (free), forced, and mixed. Natural convection occurs as a result of fluid (air) being in contact with a heated surface. The density of the fluid decreases, causing the fluid to rise and thus creating a natural circulation. Forced convection occurs when the fluid motion is induced by external sources. These external sources include fans, pumps, blowers and other fluid movers. Mixed convection occurs when natural and forced convection are both present. This is typically observed in low-velocity flows in the presence of high-powered components.

Convection heat transfer is governed by Newton's cooling law defined as

$$Q = hA (T_s - T_f) \tag{2}$$

Irrespective of the type of convective heat transfer, Equation 2 is used for its solution. What sets the three types of convection heat transfer apart is h (heat transfer coefficient). The coefficient h is obtained from empirical data. Table 2 gives ranges of h encountered in cooling of electronic components. Also, the literature is rich with research papers that show different correlations for a heat transfer coefficient. References [Kays, et al., 1980; Wirtz, et al., 1989

TABLE 1
Material Property Typically Used in Components

Material	Maximum Use	K (W/cm°C)	Application
		Metals	
Aluminum	660	2.1	Chip conductor and wire bonds
Gold	1,063	3.4	Hybrid conductor and wire bonds
Copper	1,083	3.8	Lead frame and hybrid, PWB, and conductor
Lead	327	0.3	Solder attach
Molybdenum	2,610	1.3	Cofired on ceramic conductor
Tungsten	3,380	1.5	Cofired on ceramic conductor
		Organics	
Epoxy (70% SiO_2)	170	0.002	Packaging
Epoxy glass (FR-4)	120	0.02	Multilayer board substrate
Adv. epoxy (resin only)	180	0.02	Multilayer board substrate
Triazine	250	0.002	Hybrid dielectric
BT resin (laminate)	290	0.005	Flexible substrate
Polymide	400	0.0007	Flexible substrate
Polymide	310	0.0007	Interlayer dielectric
		Inorganics	
Alumina (ceramic)	1,600	0.3	Hybrid substrates/chip carriers
Silica (fused)	1,100	0.02	Filler for molding epoxies
Silicon nitride	2,000	0.3	Candidate substrates
Aluminum nitride	1,800	3.2	Candidate substrates
Silicon carbide	2,100	2.7	Candidate substrates
Silicon	1,400	1.5	Candidate substrates
Diamond	>3,500	20.0	Candidate encapsulation
Glass-ceramic	>1,000	0.05	Candidate substrates
Beryllia	1,500	2.6	Chip carriers

Source: Azur, 1992.

and 1984; Sparrow, et al., 1982; Sridar, et al., 1990] are examples of some of these articles that may become useful to the reader.

A word of caution seems merited at this point with respect to h and its respective correlations. As mentioned, h is a coefficient and its value is obtained from empirical data. To use a given correlation, its constraints must match your specific problem. Otherwise, that particular correlation is not suited for your analysis. The second point is that convective heat transfer analysis is an iterative process. The value of h obtained from a correlation is not absolute. Therefore, the results should be verified in the overall scope of the problem.

Radiative heat transfer occurs when heat is transported by photons or electromagnetic waves. What sets radiation heat transfer apart from conduction and

TABLE 2
Heat Transfer Coefficient Value
for Different Coolants

Coolant	h, W/m²·°C
Natural Convection (plate)	
Air	5
Engine oil	37
Water	440
Forced Convection (Uo = 10 m/s)	
Air—plate length = 0.1 m	39
Air—plate length = 0.5 m	17
Boiling	
Water at 1 atm, in a container	3,000
Water at 1 atm, at peak heat flux	35,000
Film	300
Condensation	
Steam at 1 atm—on horizontal plate	9,000–25,000
Steam at 1 atm—on vertical plate	4,000–11,000
Dropwise	60,000–120,000

convection is the medium for transport and its nonlinear dependence on temperature. Radiation heat transfer requires no medium to transport the energy, and it is always present regardless of the application. However, its magnitude and degree of contribution to overall thermal transport, similar to other modes of heat transfer, is a function of temperature difference. A typical belief that radiation heat transfer can be ignored is purely a misconception. This is more so in natural convection problems, where in excess of 20% of thermal transport is attributed to radiation.

Radiation heat transfer is governed by Equation 3

$$Q = \varepsilon\sigma F_{hc} A (T_h^4 - T_c^4) \tag{3}$$

Values for F_{hc} (view or shape factor) are tabulated in any heat transfer text [Incropera, 1990; White, 1984]. A point that should be noted is that T_h and T_c are in absolute units of temperature. For more in-depth review of the fundamentals of heat transfer, the reader is referred to references [Incropera, 1990; White, 1984].

Equations 1 through 3 define the basic concepts of heat transfer, and more importantly, parameters impacting it. The common denominator in the three equations is the temperature difference, stemming from the definition of heat transfer. Second, area, either cross-sectional in the case of conduction, or surface, in the cases of radiation and convection, plays a significant role in the magnitude of heat transfer. Likewise, it can be a limiting factor in the transport of heat. In the case of convection heat transfer, heat sinks are implemented to increase surface area. Similar surface enhancements are also used for heat transfer improvement by radiation.

Material properties also play a significant role in heat transfer. Thermal conductivity and surface emissivity are such examples. Heat transfer coefficient h in Equation 2, is another parameter affecting heat transfer. Although h is not a property, it is a function of air velocity and cooling fluid properties. Depending upon a given application, fluid velocity can be the dominating parameter in influencing the magnitude of h. Hence, knowledge of the value of velocity is essential in heat transfer calculation. This is the subject of the next section.

FLUID FLOW IN CIRCUIT PACKS (PCBs)

As described in the previous section, air velocity can have a major impact on the overall heat transfer. In the following sections, the dominant role that the velocity plays becomes more apparent as I describe temperature rises of the board and component. Here, I show a methodology for determining air velocity in a circuit pack based on the integral form of the momentum equation [Murray, 1990]. The focus of this section is at the circuit pack and component levels; however, its expansion to the system level is obvious as terms are not confined to these levels.

We would like to develop an expression from which air pressure drop and, subsequently, velocity can be obtained. Consider the local momentum integral balance under the assumption of approximately equal inlet and outlet momentum fluxes:

$$\int_\sigma (\Delta P)\, dA_\sigma - A_\sigma \Sigma (K\rho V^2/2) - \int_s \tau_w\, dA_s \tag{4}$$

$$= \text{Momentum flux}_{in} - \text{Momentum flux}_{out} + \text{Buoyancy}$$

Integrating by using the first order model yields

$$\Delta P = K\rho V^2/2 + (1/A_\sigma)\, \Sigma\, (\tau_i A_{s,i}/A_\sigma), i = \text{all surfaces} \tag{5}$$

Estimating the shear stress by the following:

$$\tau_w = \mu dV/dy \sim \mu V/\delta \tag{6}$$

Where δ is the momentum boundary layer thickness. The above equation for pressure drop becomes

$$\Delta P = (D_{form} + D_{fric})\, R^2 \tag{7}$$

where

$$D_{form} = \rho K/2A_\sigma^2 \tag{8}$$

and

$$D_{fric} = (\mu/A_s^3 V^2)\, \Sigma\, \left(V_i A_{s,i}/\delta_i \right) \tag{9}$$

The form drag (D_{form}) is a function of the loss factor (K), which is based on contraction and expansion in ducts. The contraction (cc) and expansion (ec) co-efficients are obtained from the following equations:

$$K_{cc} = 0.42[1 - (A_{in}/A_{out})^2]$$ (10)

and

$$K_{cc} = [1 - (A_{in}/A_{out})^2]^2$$ (11)

The above equations are accurate as long as the inlet to outlet area ratio equals 0.76, beyond this point $K_{cc} = K_{ec}$.

The inlet area for the contractive loss is the channel height multiplied by the control volume width, while the exit area is the inlet area minus the height times the width of the component. This is reversed for the expansion loss because the inlet area is around the component while the outlet is the area of the control volume. We should note that because of typical low air velocities observed in electronic systems, frictional losses can be ignored without jeopardizing the accuracy of the solution.

Having developed an expression for pressure drop, now we need to calculate the velocity. For a forced convection system, where a fan is used, the procedure is straight forward. To obtain the velocity, we have to match the system curve to the fan curve to obtain the volumetric flow rate. Calculation of velocity for natural convection requires the following steps.

Initially, the volumetric flow rate through the system based on the ambient temperature is estimated by Equation 9:

$$R_{system} = (g\, H_{system}Q/(2C_pT_{amb}D_{sys}))\,1/3$$ (12)

Here the word "system" applies to the problem at hand. This flow estimate is based on the equality

$$\Delta P = gH_b\Delta\rho = D_{system}R_{system}^2$$ (13)

where density (ρ) is obtained from the ideal gas law and

$$D_{system} = D_{form} + D_{fric}$$ (14)

An estimate for T can be made on the basis of

$$Q = \rho\, R_{system}\, Cp(T - T_{amb})$$ (15)

After calculating the temperature changes for each board in the rack, the flow per board is calculated based on the following:

$$R_{system} = (gH\rho\Delta T/(2\, T_{in}\, D_{system}))^{0.5}$$ (16)

The preceeding process is iterative in nature, and these calculations are repeated until a converged solution is obtained.

Equations 8 and 16 show how the air velocity in a system or board can be calculated and what the pertinent parameters are that impact flow and subsequently heat transfer. These equations clearly highlight the important role which physical geometry plays in air flow through electronic circuit packs and enclosures.

HEAT TRANSFER IN ELECTRONIC COMPONENTS (MODULES)

This section is intended to familiarize the reader with generation, spreading, and eventual departure of heat from a component. It should provide the reader with sufficient understanding to be warned of simplistic solution approaches to a rather complex problem.

Electronic components (modules) are made of an aggregate of materials with different physical geometry. The central core of the component is the die containing resistors, capacitors, inductors, transistors, etc., typically bonded to a silicon. The combination of these parts form the chip where the electrical functions are performed. The chip is mounted by an epoxy or other material onto the substrate. The chips and substrate assembly is, in some cases, molded to protect the chip from potential environmental or processing hazards. Some multichip modules (MCM), or single-chip modules (SCM) continue to be made without molding, but the plastic molded packaging is becoming an industry standard.

The electrical signals are brought to the component via the leads and then to the chip(s) through the wire bonds. The leads are either press fitted or soldered to the substrate. The component is connected to the board through its leads either by surface-mounting or through-hole methods. This connection creates a package with potentially multiple heat sources in thermal communication with each other via the substrate and its ambient.

The generated heat at the chips seeks the "least resistant" path to reach the sinks. The sinks where the heat is eventually transferred are the cooling fluid and the board. The paths available for heat flow are through the molding material and the leads after it is first conducted to the substrate. The flow of heat is impeded by each material, regardless of its thickness, as it travels from the source(s) to the sinks. The wire bond also provides another avenue for the transport of heat. This can become a major path if the cross-sectional area is proportionally large.

As the heat reaches the leads, part of it is conducted to the board and the rest is either radiated and/or convected to the ambient. Also, a similar process occurs as the heat reaches the physical boundaries of the component. The heat-transfer path through the component is not necessarily one-dimensional and it tends to flow in any direction that poses the smallest resistance. Combination of multiple heat sources and different possible avenues for heat flow has created a rather complex and nonuniform temperature field in SCMs and MCMs.

HEAT TRANSFER IN CIRCUIT PACKS
AND ITS EFFECT ON COMPONENTS

The circuit board acts as a channel not only for electrical communication but a thermal one as well. Similar to electrical signals transmitted through the routings on a board, the thermal signals are transmitted through the solid. The thermal signals go through any material, even material with very low thermal conductivity. The rate of heat transmission is a function of the material conductivity and cross-sectional area. Of course, this transmission rate is different between electrical and thermal, but it occurs nevertheless. Based on this discussion, two questions are raised: what are the modes of thermal communication and how does it affect thermal design?

The heat transfer in a circuit pack takes place by fluid motion (convection) and conduction heat transfer in the board. For the fluid motion, imagine a circuit pack inserted vertically into a frame and air used as the cooling fluid. The circuit pack is populated with a number of components and there are two critical components placed on the bottom and top of the board. Remember to make these components functional; a number of other components are necessary. These components may be moderate to high in power dissipation. As the air enters the channel formed by the circuit pack and its neighboring circuit packs, its temperature rises as the result of convective heat transfer. By the time air gets to the top of the board, where the other critical component resides, its temperature can be significantly higher than the inlet. Part of the heat that was dissipated by the bottom critical component was carried to the other critical component by air. Although these components are sufficiently apart, they are in thermal contact via the fluid motion—thus, thermally coupled.

Another mode of thermal communication is through the board itself. The circuit board material can vary from pure glass–epoxy to glass–epoxy with multiple layers of copper. Because of electrical functionality, most components require several layers of copper in the board. The copper increases thermal conductivity of the board, hence enhancing conduction thermal coupling between components. The effect or radius of influence of conduction coupling is a direct function of board thermal conductivity. Let's reconsider the vertically inserted circuit pack and the two critical components. For the sake of discussion consider the board to have four layers of copper and assume we can turn these components on and off. Turn off the component in the bottom and switch on the top one while monitoring the temperature of the bottom one. We will see that the temperature (junction or case) of the one that is off varies as we switch (after steady state) the other component. If these components were placed in close proximity, the response would have been faster. Also, if the board was an eight layer instead of four then the component temperature would have been higher. If the component is low in power but surrounded by higher power components on a board, the impact of neighboring components must be considered in the thermal design [Manno, 1992]. This simplistic example illustrates the importance of thermal coupling through the circuit pack and its potential impact on thermal design.

THERMAL COUPLING
IN ELECTRONIC ENCLOSURES

To appreciate the impact of system (enclosure) on thermal performance of components, it is necessary to review the thermal phenomenon in an enclosure. The thermal phenomenon is described by coupling between components and system parts. The term "coupling" implies interdependency. Coupling means that the thermal performance of the component is directly dependent on other components and systems. As I previously discussed, the heat-transport mechanism, either by the fluid or through solid, can become very complex. To gain a better understanding of thermal coupling, we must focus on each system part. We first start from the component and go up to the frame and see how the heat is transported within an electronic system. This should assist us with the design process by helping to generate the necessary information or at least in asking the right questions.

An electronic enclosure is composed of the following:

1. Environment
2. System
3. Shelf (cage)
4. Circuit pack
5. Component
6. Die parts

Figure 1 is a schematic view of the above. What follows is an elaboration of the above list.

COMPONENT

The thermal process in a component was described in the section on Heat Transfer in Electronic Components. Since MCMs have gained more attention and are appearing in many products, a brief discussion about MCMs merits the effort. MCMs by their definition contain more than one chip. These chips are residing on one or both sides of a substrate. If we look at an MCM closely, we see that it resembles a miniature circuit pack. Many of the thermal coupling issues that we discussed in the section 2 on Heat Transfer in Circuit Packs and Its Effect as Components are also prevalent within MCMs. Since the chips reside on a substrate and substrates typically have high thermal conductivity, the potential for conduction thermal coupling is even greater than the circuit pack. Much thermal spreading takes place as a direct result of substrate thermal conductivity. This suggests that the critical chip(s) is not necessarily the one that is the largest or has the highest power dissipation. Because of the thermal spreading, smaller chips with potentially lower junction temperature tolerance may become the critical ones. Therefore, in the limit analysis, it is essential to look

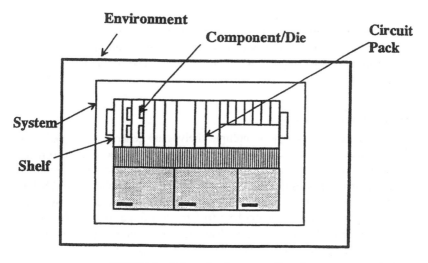

FIGURE 1.   Schematic of system configuration.

at the junction temperature of each chip and ensure that it is within its design specification.

BOARD/SHELF

The shelf or card holder (cage) is where the circuit packs reside in the system. The shelf basically acts as a housing and facilitates electrical connection of the boards through the backplane. Boards are normally inserted into the shelves through card guides. Except in some specialized cases where a latching mechanism is used to rigidly attach the board to the shelf, the boards are loosely fitted inside the shelf (e.g., a PC motherboard). In these cases, the primary holder of the circuit packs is the backplane or the motherboard. Therefore, the necessary contact to facilitate conduction heat transfer from the board to the shelf does not exist.

Heat generated at the component is conducted to the board through the leads and the gap. Heat is then transported away from the board by all three modes of heat transfer. Since poor contact exists between the board and the card guides, the conduction heat transfer is significantly less than that of convection and radiation, respectively. This may not be the case, however, if the boards are in good thermal contact with the shelf. We can generalize, based on observation, that in most system designs conduction coupling between the board and the shelf is very weak, e.g., telecommunication rack or PC motherboard.

The backplane, e.g., motherboard in a PC, is another avenue for the heat to be transported to the ambient or the shelf. If the thermal conductivity of the board is very large, i.e., multilayered boards with several layers of copper, conduction heat transfer to the backplane can be significant. Depending upon the magnitude of heat dissipation and the nature of contact between the board(s) and the backplane, boards can also be thermally coupled via the backplane. The thermal coupling by convection and radiation heat transfers is significantly larger than conduction heat transfer. Therefore, board/shelf combination provides another avenue for the heat to be coupled with the rest of the system.

FRAME (ENCLOSURE)

Frames that house single or multiple shelves are generally designed to be isolated from the shelves. The heat that is generated within the system normally finds its way out through the vent holes. Although this constitutes the bulk of heat flow, there exists significant thermal coupling between the boards/shelves and the frame. The thermal coupling, in the order of significance, is by radiation, convection, and conduction heat transfer. Since the frame is in contact with the system ambient, it can act both as a sink and source of heat for the system. In general, the coupling cannot be categorized as an insignificant part of the thermal response of the system.

The magnitude of conduction heat transfer is system-design dependent. A general statement cannot be made that heat transferred by conduction is small compared to other modes of heat transfer. But, because of the contact resistance, the conduction heat transfer between the shelves and the frame is usually weak. In addition, convective heating of the frame is also design dependent. If the flow of the coolant is in close contact with the frame, the convection heating will then be appreciably more. The radiation heat transfer, however, is generally the predominant mode of thermal coupling between the shelves and the frame. The radiation heat transfer tends to be even more significant if the system is cooled by natural convection.

ENVIRONMENT

The frame is coupled to the surrounding ambient via radiation and convection heat transfer. The system ambient can act both as a source and a sink. Convection and radiation cooling and heating is possible depending upon the nature of the system ambient, i.e., open atmosphere or climatically controlled buildings. The magnitude of the heat transfer can vary significantly with the changes in the system surroundings. This can constitute a major portion of the total energy transport to or from the system. Thus, frame-to-ambient thermal coupling must be an integral part of the thermal design consideration and limit analysis.

We can conclude that the thermal transport process in electronic systems is quite involved and can become complex. Because of many different thermal processes and strong coupling at various system levels, thermal bookkeeping is necessary for accurate analysis. In addition, it should be clear that we cannot only focus on a component (module) without considering the system, environment, and other parameters affecting thermal transport.

SYSTEM-LEVEL APPROACH TO THERMAL MANAGEMENT

I have described how system environment and parameters affecting it can impact MCM thermal performance. To remind ourselves, the objective of thermal design or analysis is to ensure component junction temperature meets design specifications. It is apparent from our discussion up to now that, in thermal design of any component, the component alone is not sufficient. Parameters affecting junction temperature vary from system ambient to the neighboring components and must be included in the design and limit analysis.

Two procedures are recommended for thermal design or analysis. First, thermal design simply requires system down approach. It implies that we have to look at the system ambient and work our way down to the component of interest. All parameters affecting the process should be considered. The second is a methodical approach to thermal design or analysis. In the previous section, I referred to this as "thermal bookkeeping." What it implies is that a successful analysis requires keeping track of all the parameters influencing design. Superficial or casual treatment of these parameters (starting from the environment to the component) will yield less than desirable results, [Okutani et al., 1984].

UNDERSTANDING THE CONCEPT OF COOLING LIMIT

We now explore why the question of air cooling limit is important to thermal engineers. Perhaps we can shed some light on this question by restating the objective of electronics cooling. In thermal design of electronic systems, our goal is to ensure that component junction temperature, either an MCM or SCM, will be retained below a certain limit. This limit by default is 125°C. I say by default because there are no references that argue for this or any limit. However, the reason for the low temperature is rather obvious. The activation energy attributing to the expected life of the component is exponentially dependent on temperature [Klinger et. al., 1990]. Further, the silicon tends to be temperature sensitive and at elevated temperatures detectable performance degradations are observed. Therefore, to ensure that the design is within the specifications, one needs to know how much heat can be removed by the cooling system of an electronic enclosure—lending itself to the concept of limit.

If we consider the discussion of the previous sections and the goal of electronics cooling, the focus of the limit becomes obvious. Typically, as will be seen throughout this chapter, any discussion with respect to a limit has been focused on the total heat removal from the channel formed by the circuit packs or the entire system. But, as I attempted to highlight, the point of contention is the component and how well it can be kept below its limit of 125°C. The maximum heat removal from the channel becomes a secondary issue. However, the process of limit analysis is more meaningful if we look at the temperature rise instead of the absolute temperature.

The earlier focus on the channel or the system as the limiting prospect has diminished the potential of air cooling for some systems. If we look at the excellent pioneering work [Kraus and Bar-Cohen, 1983], Figure (2) shows the potential component case to ambient temperature rise as a function of heat flux.

For the sake of understanding the limit consider the following: a plastic molded package with a maximum junction temperature (T_j) of 125°C, an ambient temperature of 50°C, and a junction to ambient temperature rise of 75°C. If we assume a 10° to 15°C temperature drop from junction to case, typical of most plastic molded components, case to ambient temperature rise will be 60°C. If we were to cool this component with air in natural or forced convection, the expected heat flux from Figure 2, will be 0.05 and 0.15 W/cm², respectively. Note that in this exercise we did not include any board or system level coupling and effectively looked at the component when it would reside on a glass–epoxy board.

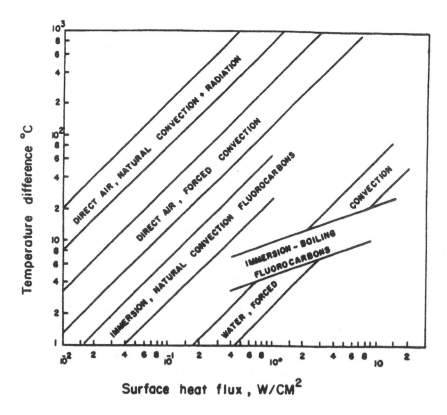

FIGURE 2. Temperature differences attainable as a function of heat fluxes for various heat transfer modes and coolant fluids [Kraus and Bar-Cohen, 1983].

Let's also consider Figure 3 [Simons, 1983], which shows the heat transfer coefficient for different fluids. Figure 3 suggests that the highest attainable value for heat transfer coefficient is 0.0025 and 0.022 for natural and forced convection using air, respectively.

If we use Equation 2 and the case-to-ambient temperature rise (approximately 60°C, used above), we get 0.125 and 1.1 W/cm^2 for natural and forced convection, respectively. Comparison of the heat fluxes obtained from Figures 2 and 3 suggests a range of heat flux capability and depicts a very limiting prospect for air cooling of electronic systems. When more than 80% of systems are cooled by air, one begins to wonder how it is done.

Consider the more recent work by the author [Azar et al., 1992] and Hilbert [Hilbert et al., 1990]. Both systems consisted of narrow channel heat sinks along with a Muffin fan as the air mover. The heat flux in the case of [Azar et al., 1992] was 20 W/cm^2 with components placed on pure glass epoxy board. This

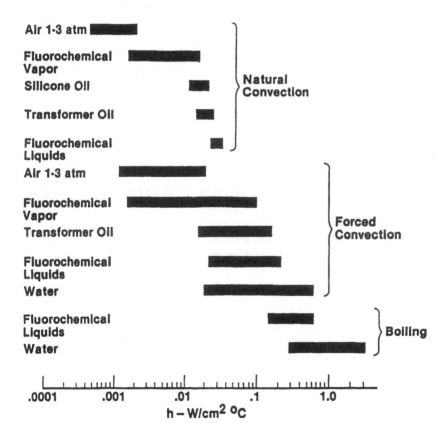

FIGURE 3.   Range of heat transfer coefficients for various coolants [Simons, 1983].

implied that there was minimal conduction heat transfer from the components to the board and the primary path of heat loss was by convection. The temperature rise in this case was 28°C, which was significantly below the limit we had discussed before. This suggests that by perhaps increasing the volumetric flow rate and convective surface area we can still increase the potential heat flux from the components. The reader can appreciate the fact that if the board was multilayer (copper ground layers), the thermal performance would have been even better because the board would have acted as a heat sink for the components.

There are a few salient points in this exercise. Foremost, the component junction temperature rise above ambient is a more realistic criterion for gauging cooling limit than the channel heat flux. Use of the heat flux, either from the component or board, does not provide a realistic picture of the capability of cooling with air. The temperature is a much better gauge which is in line with the design process. Secondly, there is limited data on the potential range of heat transfer coefficients or heat removal capability when extended surfaces (heat sinks) are used. Third, there is a need to be focused on component thermal transport when designing the cooling system. Considering the level of thermal coupling taking place in a system, as discussed in Section 3, the system-level-down approach is essential for proper gauging or design of the cooling system. However, one should note that the data from Figures 2 and 3 provide a good starting point in the design process.

PARAMETERS IMPACTING HEAT TRANSFER

In designing the cooling system, be it natural or forced convection, one typically starts with the calculation of component junction temperature for different cooling modes. The hierarchy of cooling methods considered starts with natural convection and extends into high velocity forced convection. The junction temperature is calculated for the worst-case ambient with natural convection cooling first and then extends into high-velocity forced convection. The criterion for determining suitability of a cooling mode is the following:

$$\eta = T_{j,calculated}/125 \leq 0.9 \qquad (17)$$

that allows a 10% margin. If η is not satisfied for natural convection, enhancements, such as improving radiation heat transfer or increasing convective surfaces, and others are examined. If the criterion is still not satisfied, higher convection heat transfer modes should be considered. This unsatisfied criterion suggests an unavoidable iterative process to seek a desirable cooling mode while barring the tremendous market pressures to adhere to the simplest cooling mode (i.e., natural convection).

To reduce the iterative process it is desirable for a thermal engineer to have the important parameters and their impacts to heat transfer *a priori*.

Furthermore, the order of impact or the significance of these parameters is equally desirable. Hence, we attempt to shed some light on this so-called iterative process by looking at the heat transfer from a component. To further explore the contribution or significance of each parameter, we will use order of magnitude analysis to assist us in this task.

HEAT TRANSFER IN A CHANNEL

In this section, I attempt to develop a first-order model of heat transfer in a channel formed by adjacent circuit packs. The final objective is to develop the parameters affecting heat transfer. In addition, we would like to develop an expression for the air and board temperature rise in the channel.

In most applications, circuit packs are placed parallel with each other to form a channel. Telecommunication racks and PCs are examples of these channels where both walls have a specified heat flux. In the case of even a single circuit pack in an enclosure, a channel is still formed containing a single heated wall. Consider a typical case of a channel formed by four circuit packs in a rack shown in Figure 4. We are going to focus on the channel formed by circuit packs 2 and 3 for the purpose of this development. It is apparent to the reader that the development is independent of the channel considered. Begin by stating some assumptions:

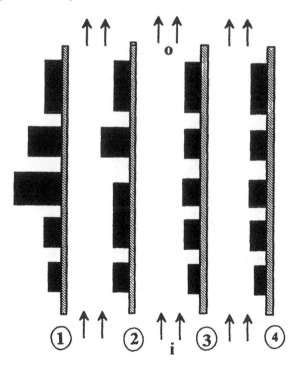

FIGURE 4.   Channel formed by adjacent circuit packs placed in a rack (shelf).

1. The heat flux is uniform over the circuit pack—this ignores the local hot spot and assumes heat is uniformly distributed. In many cases this is a valid assumption, especially when the circuit board contains multiple copper layers and components are fairly uniform in power dissipation.
2. All the heat generated leaves the board only on one side—in reality, both sides of the board dissipate the generated power. However, since we are considering a rack of boards of equal power dissipation, the overall power dissipation in the channel is the same. If the board power dissipation is not the same then the channel heat transfer can be estimated [Azar et al., 1994].
3. Radiation heat transfer between the boards is negligible since both boards are of equal temperatures.
4. Inlet channel temperature T_i and board power dissipation, Q, are given.

The heat transfer from board 3 is obtained from Equation 2

$$Q = hA(T_b - T_m) \tag{18}$$

where T_m is the mean air temperature in the channel give by

$$T_m = (T_i + T_o)/2 \tag{19}$$

Since T_o is not known we can obtain it from the following equation:

$$Q = mC_p(T_o - T_i) \tag{20}$$

This yields the air temperature rise inside the channel

$$T_o - T_i = Q/mC_p \tag{21}$$

solving for T_o

$$T_o = Q/mC_p + T_i \tag{22}$$

Substituting into Equation (19) for T_m, we get

$$T_m = 0.5(2T_i + Q/mC_p) = T_i + Q/2mC_p \tag{23}$$

Substitute equation 23 into 18 for T_m

$$Q = hA[T_b - T_i - Q/2mC_p] \tag{24}$$

Solving for the board temperature:

$$T_b = [1/hA + 1/2mC_p]Q + T_i \tag{25}$$

Solving for the board temperature rise above the channel inlet:

$$T_b - T_i = [1/hA + 1/2mC_p]Q \qquad (26)$$

The mass flow rate, m, is equal to ρ VA, where V is the air velocity. Equations 18 and 26 are applicable to natural and forced convection, and both equations display strong dependency on the air velocity. Hence, effective determination of the temperature rise requires calculation of the air velocity as shown in the section on Fluid Flow in Circuit Packs. In addition, Equation 26 also requires a value for h where it is obtained from a number of correlations in the literature [Azar, 1994].

HEAT TRANSFER FROM A COMPONENT

In this section, we will develop a first-order expression showing the heat transfer from the component. This process will yield an expression for the junction temperature as a function of parameters impacting it. Consider Figure 5 where a half of a component is shown. We will apply a control volume to different segments of the component and apply conservation of energy to each control volume. The component is a plastic molded package that is mounted on a printed wiring board—a case most commonly encountered.

We start with the molding that houses the die, wire bonds, and the chip-carrier. Apply a control volume around the molding as shown in Figure 6.

The heat transfer leaving the control volume is characterized by Q_1, Q_2, and Q_c. Q_1 and Q_2 are heat transfer from the top and bottom of the molding, respectively, and are composed of radiation (r) and convection (h). Note that Q_2 can also include the conduction heat transfer through the air gap. The important point is that there is heat transfer from the bottom side of the component. Q_c signifies the heat conduction through the lead.

$$Q_1 = q_h + q_r, \quad Q_2 = q_h' + q_r' \qquad (27)$$

Applying the conservation of energy to the control volume and assuming steady-state condition, we get the following:

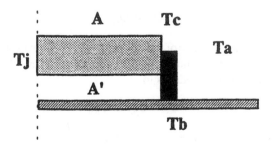

FIGURE 5. Half of a component with plastic molding.

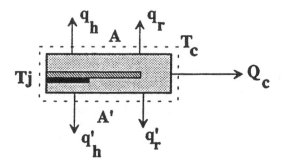

FIGURE 6.   Energy balance on the molding part.

$$Q_{in} + Q_{dissipation} = Q_{out} \tag{28}$$

Since there is no heat entering the control volume, the conservation equation becomes

$$Q_{dissipation} = q'' = Q_{out} \tag{29}$$

where $q'' = Q_1 + Q_2 + Q_c$ solving for Q_c

$$Q_c = q'' - (Q_1 + Q_2) = (kA/L)_{eff}(T_j - T_c) = R_{eff}^{-1}(T_j - T_c) \tag{30}$$

Solve for T_c a reference temperature on the component:

$$T_c = T_j - R_{eff}Q_c \tag{31}$$

The value of $(kA/L)_{eff}$ can be obtained and is subject to further discussion [Azar, 1994]. For the purpose of this discussion we should note that the physical geometry plays a significant role in the thermal performance. Performing an energy balance on the lead:

Applying conservation of energy to the lead, Figure 7, Equation 17 reduces to

$$Q_{in} = Q_{out} \tag{32}$$

The heat transfer into the control volume is by conduction; Q_c and leaving the control volume is by convection from lead surface $(Q_{h,L})$ and the board $(Q_{h,b})$. Hence, the above equation reduces to

$$Q_c = Q_{h,L} + Q_{h,b} \tag{33}$$

where $Q_{h,b}$ is the convection heat transfer from top and bottom of the board and is equal to conduction heat transfer in the board

$$Q_{h,b} = Q_{board} = k_b A/L(\Delta T)_{board} \tag{34}$$

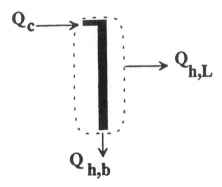

FIGURE 7. Energy balance on the lead.

Here, k_b is the board conductivity and ΔT is the temperature difference between T_b and another isotherm typically at the centerline between two adjacent components. A and L refer to the cross-sectional area and the length of the board associated with the conduction heat transfer, respectively. Hence, k_b, an important parameter, is implicitly introduced in $Q_{h,b}$. Estimating lead surface temperature as

$$T_s = (T_c + T_b)/2 \tag{35}$$

then, solving for Q_c in terms of temperatures

$$Q_c = hA_L((T_c + T_b)/2 - T_m) + hA_b(2T_b - T_{m,b} - T_m) \tag{36}$$

Solving for T_b

$$T_b = \{Q_c - hA_L T_c/2 + (hA_L + hA_b)T_m + hA_b T_{m,b}\}/(hA_L/2 + 2hA_b. \tag{37}$$

Solving for T_j, assuming h's are equal (if we have the value for h around component, we do not need to make this assumption)

$$T_j = T_c + R_{eff}\{T_b(hA_L/2 + 2hA_b) + hA_L T_c/2 \\ - h(A_L + A_b)T_m - hA_b T_{m,b}\} \tag{38}$$

The above expression requires T_c and T_b to be known. Considering that leads tend to be highly conductive in nature, one would not expect a large temperature drop across the lead. Hence, assuming T_c to be equal to T_b will not significantly jeopardize the accuracy of the solution, thus eliminating one variable.

Equation 23 requires $T_{m,b}$, which is the mean air temperature of the channel formed by boards 3 and 4 in Figure 4.

$$T_{m,b} = (T_o + T_i)_{3,4}/2 = (Q_4/2mC_p) + T_i \tag{39}$$

Substituting the equivalent of flow rate yields

$$T_{m,b} = (Q_4 / (2\rho V_{3,4} A_{3,4} C_p)) + T_i \qquad (40)$$

Substituting Equation 40 into equation 38, and setting $T_c = T_b$, we get an expression for T_j as a function of parameters that are readily measurable.

$$T_j = T_b + R_{eff} \{ T_b (hA_L / 2 + 2hA_b) + hA_L T_b / 2$$
$$- h(A_L + A_b) T_m - hA_b (Q_4 / (2\rho V_{3,4} A_{3,4} C_p)) + T_i \} \qquad (41)$$

Substitution of the values of T_m and $T_{m,b}$ in terms of the flow rate and power dissipation results in the following:

$$T_j - T_i = (T_b - T_i) + R_{eff} \{ hA_L (T_b - T_i) - hA_L Q_3 / m_{2,3} C_p \qquad (42)$$
$$+ 2hA_b / C_p [(Q_3/m)_{2,3} + (Q_4 / m)_{3,4}] \}$$

Equation 41 or 42 represents the junction temperature of a component residing on a board. This model is for the case when the power dissipation of the component being modeled is higher than its neighbors. The model explicitly does not include heat transfer by conduction through the board from the adjacent components. If this is not the case, the model shown here can be expanded to include the conduction heat transfer between the component and its neighbor. This is done at the lead level, Figure 7, where additional heat transfer enters the control volume at the board.

The reader also can appreciate the sensitivity of the model to R_{eff}—component internal thermal resistance. An accurate estimation of this parameter can improve in the level of accuracy in T_j.

Equation 41 can be used irrespective of the mode of convection heat transfer. The junction temperature can be obtained if the value of heat transfer coefficient (h) is known. The reader may note that the process for obtaining the junction temperature is indeed iterative. Solution of a set of Equations 7, 23, 25 and 42 is required in order to obtain T_j.

SIGNIFICANT PARAMETERS

The junction temperature, Equation 41, is the parameter of utmost interest to thermal engineers in electronics cooling. We would like to know what the parameters are that have the most impact on the value of T_j. We developed a series of equations that relate junction temperature to other parameters affecting it. Our objective in thermal management is to ensure that T_j is as small as possible within given constraints. By examining equation 41, maybe we can shed some light on the important parameters and gain some insight as to how we can reduce T_j.

We can reformulate equation 36 by the following:

$$T_j = f(A_L, A_b, V, T_m, T_i, R_{eff}) \qquad (43)$$

Or expressed in general terms:

$$T_j = f\,(\text{geometry}, \text{properties}, \text{velocity}) \tag{44}$$

Equation 27 effectively shows the parameters impacting T_j. For T_j to be as small as possible (our goal in thermal management), the terms in the braces must be minimized.

The parameter that has a dominant presence in Equation 41 is the air velocity. This is seen in the convection terms associated with the lead surface and board areas. Heat transfer coefficient (h) is directly proportional to velocity and to change of temperature (density) in forced and natural convection systems, respectively. This is somewhat intuitive and obvious since the ultimate sink for heat is the air stream. It is interesting to see that convection heat transfer from the head plays such a role in the magnitude of T_j. This is often overlooked in modeling and design of electronic components and circuit packs.

In addition, we see the presence of T_i—inlet temperature to the channel. This stems directly from our earlier discussion, Section 3, that in thermal analysis system down approach is the only way to obtain an accurate solution. This is evident in Equation 41, since the value for T_i can be obtained by system level analysis or direct measurement.

We also see the impact of neighboring boards in the value of T_j. The adjacent boards impact the mean air temperature (T_m) in the channels (front and back) subsequently affecting heat transfer from the board. Further, the adjacent board geometry (layout) also impacts heat transfer. This is observed in the hA_b ($Q_4/(2\rho V_{3,4}A_{3,4}C_p)$) term of Equation 41 by the presence of velocity in the denominator. The resistance to flow, stemming from component layout, will directly impact flow inside the channel, Equations 7–9.

The role that a board plays in thermal performance of the component cannot be emphasized enough. Boards with higher thermal conductivity Equation 34 can be instrumental in removing heat from the component. Higher board conductivity can increase A_b, resulting in higher convection heat transfer from the board. Again, this is demonstrated by equation 41, even though board thermal conductivity does not directly appear in the equation.

Irrespective of mode of convection heat transfer, I mentioned the important role velocity plays in the magnitude of T_j. An often overlooked point in circuit board design is the board layout. Board layout plays a pivotal role in component flow exposure. Poorly laid-out boards can create zones of stagnation that hamper convection heat transfer from the board and component. Azar and Russell (1991) provide many flow visualizations of such effects. This can also be deduced from Equations 7–11. The pressure drop is the direct function of the expansions and contractions created by the components on the board, hence, affecting air velocity and flow distribution in the rack. The pressure drop in the system can be reduced by proper placement of components on the board. This is done such that the critical components and/or boards have the least resistance to air flow.

POTENTIAL LIMITS OF NATURAL
AND FORCED CONVECTION COOLING

The word limit conveys a message that there is a number associated with power dissipation beyond which cooling with natural and/or forced convection is not possible. In the foregoing discussions I have attempted to replace this notion with the need to look at each case explicitly. I also have advanced the argument of reshaping our thought by looking at the junction temperature limit instead of heat transfer when dealing with the issue of limit. For example, a common question we as thermal engineers face is "If this board will be dissipating x Watts, can I cool it with natural convection or do I have to use a fan?" It is convenient to have a number, say 25W per board, to compare against and respond accordingly. If we look at the question more closely, the issue is really the junction temperature limit and not the total power dissipation. Further, as I indicated in the previous section, there are numerous parameters that can impact this evaluation, e.g., channel dimensions, board material, component layout, system configuration, environment, etc. Hence, we need to look at the concept of limit from temperature perspective instead of the heat flux number that may not be applicable and is a function of several system-dependent parameters.

I need to side step and add that by no means am I suggesting that we cool all systems with natural and forced convection. The consideration for a cooling system is driven not only by physical and thermal limits but by market limits as well. A case in point is the implementation of jet impingement for cooling of the processors in a PC. The fan noise is difficult enough to bear. Imagine contending with a whistling noise coming out of your PC if it contained jet impingement. Likewise, a consumer electronics product, e.g., CD player, has its market-defined limits or a super computer, e.g., Cray, cannot have a jet turbine to provide enough flow for air cooling. Hence, in dealing with the issue of limits, other parameters in addition to junction temperature should be considered.

To address the issue of limit, I will review the work available in the literature and then focus on the temperature limit versus heat flux limit that I have advocated here. Because of the diversity in electronic systems and the above discussion on system and application dependency, I have purposefully avoided generating numbers that can stand as a limit for comparison. However, the tools for calculating the potential limits for your specific problem are provided that readers can use for their analysis.

FORCED CONVECTION LIMIT ANALYSIS

Based on the issues we raised, the literature is very limited on convection limits. The works, by Hannemann (1990), Jacobs (1989, 1990), and (Chung, 1987) are noteworthy for review.

Hannemann looked at the limits of forced convection as a function of number of chips per unit area versus number of pins per chip for the mini- and microcomputer. The analysis was done for the case when the chips were attached to a heat sink as shown in Figure 8.

The chip temperature rise was described by

$$\Delta T = \frac{\Psi_1 Q}{mC_p} + \Psi_2 R_{da} q''$$ (45)

Considering that the convective resistance is the dominant factor, the above equation reduces to

$$\Delta T = \frac{qA_C}{hA}$$ (46)

The number of chips per unit area is given by

$$N/L^2 = 1/A_b$$ (47)

substituting into Equation 27, we obtain the following:

$$\Delta T = \frac{qA_C}{h\gamma A_b}$$ (48)

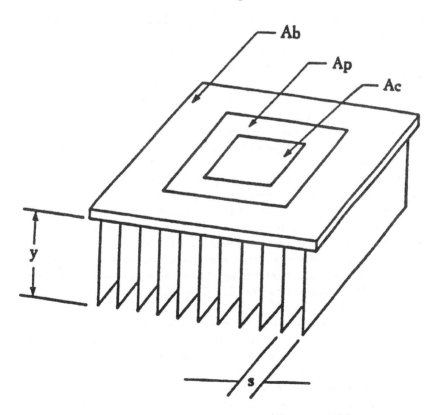

FIGURE 8. Extended surface chip cooling, [Hanneman, 1990].

where

$$\gamma = A / A_b = 1 + 2y / s, \tag{49}$$

and

$$A = A_b + 2y l_f [l_f /(l_f + 1)] \tag{50}$$

Extracting the chip area that may be cooled with an extended surface area, A_b and chip heat flux q is given by

$$A_c = h\gamma\Delta T A_b / q = \frac{h\Delta T A_b}{q}(1 + \frac{6.39}{l_f^{0.5}}) \tag{51}$$

where the heat transfer coefficient is given by

$$h = 3.78(V / l_f^{0.5}) \tag{52}$$

Since $l_f^2 = A_b$ and combining Equations 50 and 51, we get

$$A_c = (8.45 A_b^{0.75} + 54 A_b^{0.5})\Delta T / q \tag{53}$$

If we set the temperature rise to 50 K, and vary the heat flux to 10 and 20 W/cm² we get Figures 9 and 10

The value of the heat flux, length of the heat sink, y, and velocity can be changed to develop different curves. The figures show the maximum thermally allowed chip density at a given pin count. The model was generated under the assumption that the internal thermal resistance is negligible and the convective resistance is the only one. Therefore, it advances the best scenario since the current technology is limited and does not allow for this situation to exist. However, the data and methodology are helpful in creating potential bounds on the problem.

There are a couple of salient points that merit mentioning. Generalization of problem-specific issues can create false information—alluding to the point of generating a "number" as yard stick for determining a limit. As Hannemann carefully pointed out, the model developed is for a very specific application dealing with chips placed on an extended surface. The second point is the dependency of the model on physical parameters and heat transfer coefficient (velocity). Both of these parameters can significantly alter the results if their values are changed as shown by Equations 51–53, further substantiating the need to look at each case individually.

Jacobs (1989, 1990) advances the application of Reynolds' analogy in electronics cooling. He suggests the use of the analogy both for forced and natural convection cases to determine the limit. Reynolds' analogy is widely used in the heat exchanger industry and Jacob's method is an interesting approach in this concept for electronics equipment.

Consider the forced convection case where the equipment is constrained by the acoustic noise limit and a maximum component-to-air temperature rise—

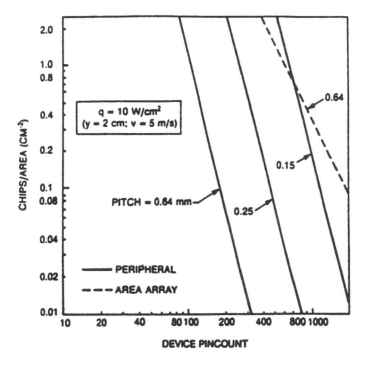

FIGURE 9. Thermal limits – q = 10 W/cm² [Hanneman, 1990].

a situation commonly encountered. Figure 11 shows a general schematic of a system that contains fan, filter and the equipment.

The noise generated in the system is proportional to the volumetric flow rate (R) and generated pressure (P)

$$R \, P^2 = \Lambda \tag{54}$$

Based on Equation 4, pressure drop as a result of physical constraints is a function of the flow resistance (D) and the volumetric flow rate. Reynolds analogy simply states that the drag force created as the result of fluid passing a heated rigid body is proportional to its heat transfer. This is expressed by the friction coefficient as a function of other nondimensional numbers:

$$C_f = 2Nu \, / (Re \, Pr^{1/3}) \tag{55}$$

If the top area of the shelf shown in Figure 11 is A_{top}, the volumetric flow rate is defined by

$$R = V \, A_{top} \tag{56}$$

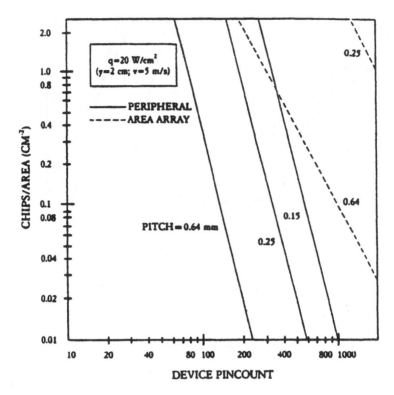

FIGURE 10. Thermal limits – q = 20 W/cm², [Hanneman, 1990].

and air flowing in a circuit pack channel at a velocity V and surface area, A_{board}, the pressure drop created is

$$P = C_f \rho V^2 (n A_{board} / A_{top}) \tag{57}$$

where n is the number of circuit cards in the rack. The pressure drop can be expressed in terms of internal and external resistances, which is given by the following equation:

$$P = \beta R^2 + DR^2 \tag{58}$$

Defining β as P/R^2 we get

$$\beta = \frac{C_f \rho n A_{board}}{A_{top}^3} \tag{59}$$

The heat transfer from two sides of a board is

$$Q = 2n A_{board} h \Delta T_{sa} \tag{60}$$

and the heat transfer coefficient is deduced from Equations 55 and 60:

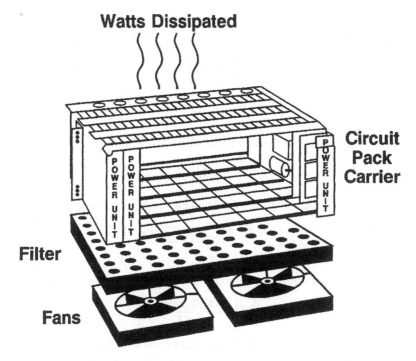

FIGURE 11. Schematic of a shelf with filter and fan [Jacobs, 1989].

$$h = \xi C_f RkPr^{1/3} / 2VA_{top} \tag{61}$$

where $\xi = 0.7$ is the calibration factor to fit the Reynolds analogy to the circuit pack problem. Solving for R from Equations 54 and 58 results in

$$R = \left[\frac{\Lambda}{(\beta + D)^2}\right]^{1/5} \tag{62}$$

By successive substitution of equation 62 into 61 to eliminate R and equation 61 into 60, we can obtain an expression for maximum power dissipation:

$$Q = \frac{C_f \xi k (Pr)^{1/3} \Phi \Delta T_{sa} n A_{board}}{\upsilon A_{top}} \left[\frac{\Lambda_{max}}{(\beta + D)^2}\right] \tag{63}$$

In equation 63, Φ corresponds to the percentage of the area of the circuit board that is heated. Jacobs suggests a value equaling 0.02 for the friction coefficient (C_f). This result can also be obtained graphically if one plots system flow resistance and acoustic noise on the pressure vs. flow rate coordinate systems. The point of intersection of these two curves correspond to the maximum power dissipation.

The derivation for natural convection is similar to the above. The pressure drop created by air flow across a rack of cards was given by Equation 57 and the heat transfer from the cards by Equation 60. Note that Equation 60 can be corrected by a factor of Φ introduced earlier. Further manipulation of equation 60 by using pressure drop and friction coefficient yields

$$Q = P/R\left[\Delta T_{3a}\Phi k\frac{\xi(Pr)^{1/3}A_{top}^2}{\upsilon\rho}\right] \tag{64}$$

The air temperature rise through a rack of cards is given

$$\Delta T_{air} = \frac{Q}{R\rho C_p} \leq \Delta T_{max} \tag{65}$$

by substituting Equation 65 into Equation 64, maximum dissipation in watts is given by

$$Q = A_{top}\left[\frac{PC_p\Delta T_{max}\Delta T_{sa}\Phi\xi k(PR)^{1/3}}{\upsilon}\right]^{1/2} \tag{66}$$

and the associated volumetric flow rate (R in m³/s) is given by Equation 67.

$$R = \frac{A_{top}}{\rho}\left[\frac{P\Delta T_{sa}\Phi\xi k(Pr)^{1/3}}{vC_p\Delta T_{max}}\right]^{1/2} \tag{66}$$

Equations 66 and 67 require that the maximum temperature rise be specified and the pressure drop be given. An interesting point of Equation 67 is its independence from the number of circuit packs in the rack. At a first glance, one may think that the equations are independent from the channel height, i.e., circuit pack height. This parameter, an essential element in natural convection flows, appears inherently in the expression for pressure.

Again, there are a few points that merit a discussion. Equations 63 and 66 are based on Reynolds analogy whose foundation has constraints. The analogy, although useful, is for cases when the Pr = 1, flow is laminar and the walls are smooth. The issue of the Prandtl number can be argued successfully, since for most air-cooled applications, Pr = 0.72. However, the other constraints may create some doubt in the utility of this approach or the equations.

Smooth walls and laminar flow are not common occurrences in most electronic equipment. Then the question of utility of these equations needs to be answered. The response can be formulated in several envelops. Reynolds' analogy has been successfully applied to many heat exchanger designs that may have not agreed with the constraints. Hence, the rigidity of the constraint is questionable, since many successful applications are observed and the analysis was in satisfactory agreement with the data of Bird (1960), Bejan (1984), and Cheremisioff (1984). The literature also suggests that a more suitable relation for geometry that is more representative of channels formed by electronic circuit packs is Colburn analogy:

$$St\,Pr^{2/3} = C_f/2 \tag{68}$$

The derivation of maximum dissipation based on Equation 68 is identical to that shown so far.

The Reynolds analogy as shown here has a strong utility in the design process. The outlined methodology can certainly be used for establishing bounds on the problem before proceeding with the detailed analysis. It also provides some guidelines to the direction of the design. Hence, its usefulness, despite its limitation, should not be underestimated by the reader.

Similar to what we saw with Hannemann's approach, although methodology suggested by Jacobs appear to be more general in nature, it still requires system-specific data. This again highlights my earlier argument of avoiding a yard stick type of approach for gauging air cooling limits. System configuration and operation requirements are as diverse as the designers who make them, and these are the limits that a designer needs to consider in evaluating air cooling. This discussion may become more clear if we look at natural convection more closely.

NATURAL CONVECTION OPTIMIZATION

In attempting to maximize heat transfer from the board in a natural convection cooled system, Chung (1987) developed a series of equations based on available heat transfer coefficient correlation for the following conditions:

1. Symmetric and asymmetric isothermal boards
2. Symmetric and asymmetric isoflux boards

The intent was to obtain the maximum value for the channel spacing (board to board) beyond which air temperature rise will remain unchanged.

The convective heat transfer is given by the following:

$$Q = hA_{total} \, \Delta T = [Nu \, (k/b)] \, [2LSn] \Delta T \tag{69}$$

where A_{total} is the total surface area of the boards and $n = W/(b + d)$. The other parameters are defined in Figure 12.

Table 3 shows the optimum spacing the boards can have for the given temperature or heat flux condition.

In table 3

$$P = \frac{g\beta\rho^2}{\mu^2} \Pr \frac{\Delta T}{L} \tag{70}$$

and

$$R = \frac{g\beta\rho^2}{\mu^2} \Pr \frac{q'}{\lambda L} \tag{71}$$

In Equations 70 and 71, L is the height of the circuit board, q' is the heat flux, and $\lambda = 6.17 \times 10^{-4} + 1.57 \times 10^{-6} \, T_1$. The results of the above are shown in Figures 13 and 14.

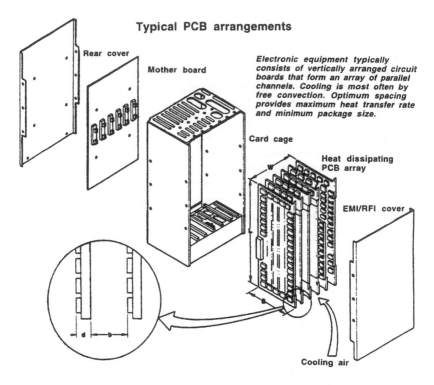

Typical PCB arrangements

Electronic equipment typically consists of vertically arranged circuit boards that form an array of parallel channels. Cooling is most often by free convection. Optimum spacing provides maximum heat transfer rate and minimum package size.

FIGURE 12. Typical PCB arrangement [Chung, 1987]. Reprinted with permission from Machine Design, March 1987, a Penton Publication.

TABLE 3
Board-to-Board Optimum Spacing
for Natural Convection Cooled System

Condition	Optimum Spacing
Symmetrically isothermal	$2b + 3d - 0.005P^{1.5}b^7 = 0$
Asymmetrically isothermal	$2b + 3d - 0.005P^{1.5}b^7 = 0$
Symmetrically isoflux	$b + 3d - 0.3133R^{0.6}b^4 = 0$
Asymmetrically isoflux	$b + 3d - 0.626R^{0.6}b^4 = 0$

The analysis presented by Chung shows the effect of heating and channel width on thermal response of the system. The analysis reveals the optimum board spacing for maximum heat transfer from the board. However, to further highlight these effects consider the work of Saxena.

Saxena (1981) highlighted the impact of other parameters that influence board or air temperature rise in a natural convection cooled system. He con-

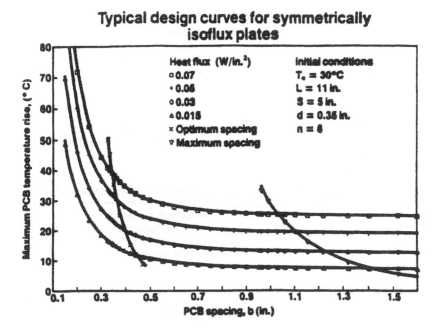

FIGURE 13. Maximum temperature rise vs. channel width for a isoflux system cooled by natural convection [Chung, 1987]. Reprinted with permission from Machine Design, March 1987, a Penton Publication.

sidered channel height, board spacing, and circuit card power dissipation in a channel. Each circuit pack was 20 cm high by 34 cm deep. The channel height was increased in increments of 20 cm by stacking the boards above each other. Figures 15 through 18 show the impact of these parameters.

These figures are self-explanatory and require very little discussion. The reader should note the variations the data show when parameters such as board spacing or power dissipation are varied. The impact of channel height shown in Figures 15–18 also are noteworthy.

In our earlier discussions, I raised the issue of component placement and its impact in overall cooling or component response. Note the work shown by Saxena and Chung does not include the impact of component placement. Another parameter to consider was work reported by Lee (1994). He considered the natural convection heat transfer for an array of parallel plates with unheated entry and exit. One can interpret that as component placement—do you place the hot components near the entrance or exit of the channel? Since the study was of a numerical nature and the data reported are nondimensional numbers the discussion may become lengthy, so I will only reflect on his observations. Lee concluded that:

1. Unheated exit has a higher fluid draw resulting in a higher heat transfer coefficient.

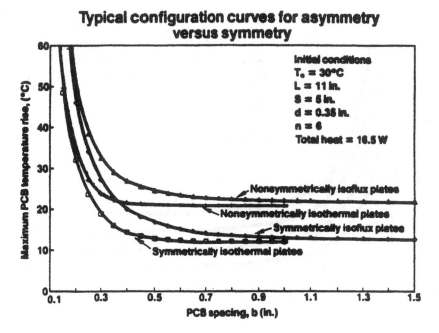

FIGURE 14. Effect of symmetry and nonsymmetry on thermal response of natural convection cooled system [Chung, 1987]. Reprinted with permission from Machine Design, March 1987, a Penton Publication.

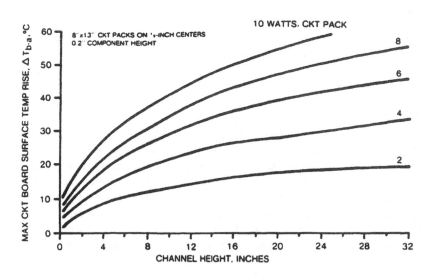

FIGURE 15. Maximum board surface temperature rise vs. channel height for 1.3 cm board spacing at different power dissipation [Saxena, 1981].

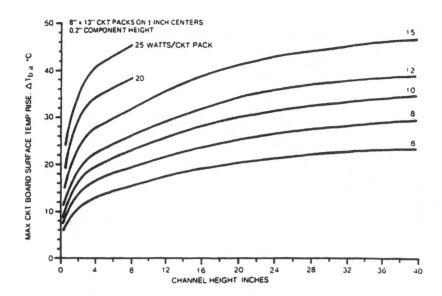

FIGURE 16. Maximum board surface temperature rise vs. channel height for 2.5 cm board spacing at different power dissipation [Saxena, 1981].

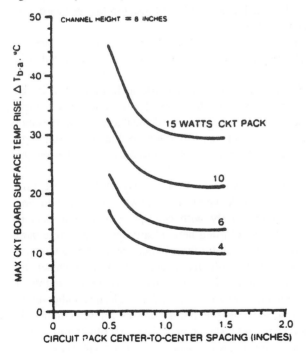

FIGURE 17. Effect of circuit pack spacing on board surface temperature rise for 20 cm channel height at different power dissipation [Saxena, 1981].

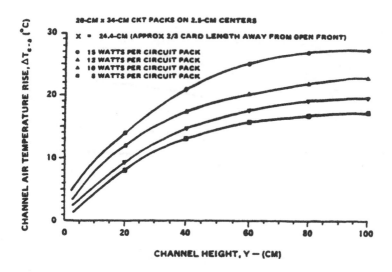

FIGURE 18. Circuit pack channel air temperature rise as a function of channel heights at different power dissipation levels [Saxena, 1981].

2. For uniform heat flux, unheated entry will result in higher board temperature than the unheated exit.
3. Effects of unheated entry or exit on heat transfer characteristics are significant, especially for the case of uniform heat flux and unheated exit.

Although the actual data may not be of immediate application in the design, the observations are definitely noteworthy. They clearly suggest that component placement can play a significant role in the thermal response of the channel. Although Lee dealt with flat plates, I used the word thermal to highlight the heat transfer and fluid flow effects that component placement has on heat transfer.

Let's revisit why we looked at natural convection. My intention was to raise the flag on the use of the yard stick type of concept when we deal with electronics cooling problems—especially with concept of limit. The data by Saxena and Chung, and others not cited here, clearly show how these parameters can influence, say, air temperature rise in the channel, board temperature rise, or heat dissipation. We should also note that the data shown here are very much system specific, and the effect of system configuration and its interaction with the environment is inherent in the data—adding another uncertainty to the problem. As a side note, the data are very useful for bounding the solution as we did with other works shown here. The work of Lee further corroborates the inadequacies of using a yard stick type of concept. Hence, once confronted with this situation, my recommendation is the grounds-up approach—integral approach.

METHODICAL PROCEDURE—AN INTEGRAL APPROACH

In the last two sections, I showed the difficulty and perhaps danger of establishing a limit for measuring the capability of cooling with convection heat transfer. Data, if available, that is based on a specific system may have been developed for another industry segment. This data may not have universal applicability and will not be suitable to telecommunications or consumer electronics. Furthermore, based on experience, we as designers are typically constrained by the temperature limit and not by how much heat a board can dissipate. The thermal engineer typically gets the design, if lucky, when the design is established and an estimate of its power is available. One never sees the reverse—that you can only dissipate x-Watts on a board—so let's design the electronics that meet the power limit (wouldn't that be a luxury). Regardless, the question "whether I can cool a given system by forced or natural convection" still lingers. Because of the diversity in design and system application, I suggest a ground-up approach that I call—integral. This approach simply states that we should:

- Consider each problem individually
- Develop a model for the junction (or critical point, e.g., lead) temperature of the components
- Determine whether it meets the temperature criterion

The reader can appreciate that for a given class of systems, e.g., PCs or telecommunication equipment, limits that can be used as a guideline may evolve. Nevertheless, the need to look at each problem individually, even in a single industry category, becomes a necessity—as some of us in the field have experienced.

How does the integral approach work? In the sections on. . . . I proposed a model for fluid flow and heat transfer in the circuit pack. The model resulted in an expression for junction temperature as a function of parameters impacting it. The intent is to look at the junction temperature of the component(s) or the temperature rise inside the channel to see whether it satisfies the design criteria. This was done by the following equation, as was discussed in the section on heat transfer from a component.

$$T_j - T_i = (T_b - T_i) + R_{eff}\{hA_L(T_b - T_i) - hA_LQ_3/m_{2,3}C_p \qquad (70)$$
$$+ 2hA_b/C_p[(Q_3/m)_{2,3} + (Q_4/m)_{3,4}]\}$$

For example, many power supplies that reside on the boards have an ambient temperature limit of approximately 75°C (50°C less than the 125°C for the ICs). Hence, the designer needs to examine what the power supply's ambient temperature might be for the worst environmental ambient—typically 49°C, the sections on. . . . Then a model, such as the one suggested above, is used to determine the "junction" temperature (typically a location on the side of the device in the case of the power supply). The junction temperature rise above ambient, obtained from the analysis, is compared with the imposed temperature limit, η, Equation 11. The values of heat transfer coefficient set the cooling mode—natural or forced convection. If η is satisfied then the cooling mode selected can be implemented. If the forced convection results based on this analysis do not satisfy η then higher cooling modes must be considered or the system needs to be redesigned.

As a side note, the author is not suggesting that the equations shown in this chapter are the sole model for these studies. The emphasis is on the approach rather than the equations. Nevertheless, the equations shown here are based on the laws of physics and is void of heuristic knowledge or system-specific parameters. The reader can tailor these equations to his/her specific problem without jeopardizing the accuracy of the results or the integrity of the approach.

In these analyses, the design temperature limits play a pivotal role. These are temperature limits that are set by standardization bodies (Bellcore or UL), market specifications, customer requirement, or any such entity. For example, in the telecommunication equipment that are used in outside enclosed cabinets, the air temperature rise in the circuit pack channel cannot exceed 10°C, or the junction temperature of the ICs cannot exceed 125°C. In most military applications, this limit is 100°C. For optical components, the case temperature is the gauge and is set at, typically, 85°C. All these limits are for the worst ambient of 49°C. As a point to note: The highest air temperature recorded on Earth was 115°F, at the city of Abadan, Iran. Hence, limit or cooling system qualification by the integral method requires temperature limit specifications.

Combinations of modeling and design temperature limits can provide us with the answers to the question of whether natural or forced convection by air can be used for cooling. We should also note that the modeling technique suggested here can also address the use of extended surfaces (heat sinks) typically used to reduce the junction temperature. The model allows for parameters such as heat transfer coefficient, areas or Q_1 and Q_2, section 5.2, to be altered such that they include either the heat sink or the effect of the heat sink. A typical example is to upgrade the value of heat transfer coefficient such that the effect of heat sink is included. Hence, the suitability of the cooling system can be determined by the use of a set of algebraic equations without the need to adhere to any unsubstantiated yard sticks.

CONCLUSION

The desire to cool with air continues to persist because of readily available and its relative ease of use, e.g., natural or forced convection. Hence, engineers in the electronics industry are continuously challenged with the question of whether a system can be cooled by the simplest mode of air cooling—natural convection. Therefore, the desire for a quick and readily available answer to this question continues to persist. The answer to this question is often sought by searching the literature for similar work.

Seeking a solution from the literature may be fruitless. When we take a cursory look at the electronic products in the market it clearly reminds the reader of the product diversity in the electronic industry. This diversity has resulted in many excellent thermal works that are industry or product specific, thus, difficult, if not impossible, to generalize. This difficulty stems from the design and application variations that electronics products face. The use of these types of information without understanding its domain of development and application can result in major flaws with a costly associated fix.

Consequently, in this chapter, a simple methodology based on conservation laws are advocated, and the use of yardsticks or presolved solutions is strongly discouraged. The methodology suggests using the integral approach to calculate the junction temperature and then use this temperature as the parameter to judge design integrity versus "heat flux in a given volume," which is typically used. The methodology emphasizes the need to obtain and understand the temperature constraints that govern one's problem. These constraints can be imposed by standardization agencies or the end user. This approach allows the engineer to analyze the problem based on its own merits and, hence, is void of any assumptions that may not be applicable to the specific problem at hand. This process empowers the engineer to design based on the governing physics and eliminates the need to use vague industry-specific yardsticks or literature-based data that may not be applicable to the problem at hand.

APPENDIX A: NOMENCLATURE

A-   area
A_b-   component thermal foot print, twice component planar area
A_c chip area
A_s surface area parallel to the stream
A_σ open flow area at the channel entrance
Cf   friction factor
C_p specific heat at constant pressure
D   drag force
F_{hc} view factor
g   gravity
H   height
h   heat transfer coefficient
k   thermal conductivity
K   loss factor
L-   half length of the component (due to symmetry)
m   mass flow rate
n   number of circuit packs

Nu	Nusselt number
P	pressure
Pr	Prandtl number
q''-	component power dissipation
Q	heat transfer rate
Q_c-	conduction heat transfer through the lead
$Q_{h,L}$-	convection heat transfer from the lead
$Q_{h,b}$-	convection heat transfer from the board
R	volumetric flow rate
Re	Reynolds number
St	Stanton number
T	temperature
T_j-	junction temperature
T_s-	lead temperature
T_{amb}-	component ambient temperature
T_b-	board or base temperature
T_c-	component reference surface temperature
T_f	fluid temperature
T_m	mean temperature
$T_{m,b}$-	mean air temperature on non-component side
T_s	lead surface temperature
$T_{s,L}$-	lead temperature
V	velocity

β	coefficient of thermal expansion
δ	boundary layer thickness
ε	emissivity
μ	viscosity
τ_w	shear stress
ρ	fluid density
σ	Boltzman constant
Ψ	convection constants defined by [Hannemann, 1990]
Λ	fan acoustic noise limit

SUBSCRIPTS

amb	ambient
b	board
c	cold
h	hot
in	inlet
L	lead
out	outlet
sa	surface to ambient
top	top open area of the shelf

REFERENCES

Azar, K. 1992. Thermal design considerations with applications to multichip modules, in *Multichip Module Technology and Alternatives—The Basic Approach*, Van Nostrand Reinhold, New York, chap. 12.

Azar, K. 1994. *Electronics Cooling—Theory and Application, short course.* Class Notes.

Azar, K., Mcleod, R.S., and Caron, R.E. 1992. Narrow channel heat sink for cooling of high powered electronic components, *IEEE Semiconductor Temp. and Thermal Manage. Symp.*, Austin, TX.

Azar, K., Pan, S.S., Parry, J., and Rosten, H. 1994. Effect of circuit pack parameters on thermal performance of electronic components in natural convection cooling, in *Proc. 10th Annu. IEEE SEMI-THERM Symp.*, San Jose, CA.

Azar, K. and Russell, E.T. 1991. Effect of component layout and geometry on the flow distribution in electronic circuit packs, *ASME J. Electron. Packag.*, 114: 50–57, March.

Bejan, A. 1984. *Convection Heat Transfer*, John Wiley & Sons, New York.

Bird, R.B., Stewart, W.E., and Lightfoot, E.N. 1960. *Transport Phenomenon*, John Wiley & Sons, New York.

Cheremisioff, N.P. 1984. *Heat Transfer Handbook*, Gulf Publishing.

Chung, J. 1987. Maximizing heat transfer from PCBs, *Machine Design*, 87–92, March 26.

Hannemann, R. 1990. Physical technology and the air cooling interconnection limits for mini- and microcomputers, *Advances for Thermal Modeling of Electronic Components and Systems*, ASME Press, 1–40.

Hilbert, C., Sommerfelt, S., Gupta, O., and Derrell, D.J. 1990. High performance micro-channel air cooling, in *Proc. 6th Annu IEEE SEMI-THERM Symp.*, 108–113, Scottsdale, AZ.

Incropera, F.P. and Dewitt, D.P. 1990. *Introduction to Heat Transfer*, Wiley and Sons, New York.

Jacobs, M.E. 1989. The practical limits of forced air-cooling of electronic equipment, *Applied Power Electron. Conf.*

Jacobs, M.E. 1990. Application of reynolds analogy to the estimation of the limits of cooling by natural convection, *Applied Power Electron. Conf.*, 431–437.

Kays, W.M. and Crawford, M.E. 1980. Convective heat and mass transfer, 2nd ed., McGraw Hill, New York, 139.

Klinger, D.J. Nakada, Y., and Menendez, M.A. 1990. *AT&T Reliability Handbook*, Van Nostrand Reinhold, New York.

Kraus, A.D. and Bar-Cohen, A. 1983. *Thermal Analysis and Control of Electronic Equipment*, McGraw-Hill, New York.

Lee, K.T. 1994. Natural convection in vertical parallel plates with an unheated entry or unheated exit. *J. Numerical Heat Transfer*, Part A, 25:477–493.

Manno, V.P. and Azar, K. 1992. The effect of neighboring components on thermal performance of air-cooled circuit packs, *ASME J. of Electron. Packag.*, 113:50–57, March.

Murray, J.C. 1990. A Thermal Model of Parallel Circuit Board In Electronic Enclosures, Master's thesis, Tufts University, Boston, MA.

Network Equipment Building System, TR-NWT-000063, Sept. 1993.

Okutani, K., et al. 1984. Packaging design of SiC ceramic multi-chip RAM module, in *Proc. of Inter. Electron. Packag. Soc.*, 299–304.

Saxena, L.S. 1981. Electronic equipment frames in free convection, in *Proc. 2nd Bell Labs Thermal Design Forum.*

Simons, R.E. 1983. Thermal management of electronic packages, *Solid State Technol.*, Oct.

Sparrow, E.M. Neithammer, J.E., and Chaboki, A. 1982. Heat transfer and pressure drop characteristics of arrays of rectangular modules encountered in electronic equipment, *Inter. J. of Heat and Mass Transfer*, 25 (7):961–973.

Sridar, S., et al. 1990. Heat transfer behavior including thermal wake effects in forced air cooling of rectangular blocks, *ASME HTD.*, WAM, (Dallas TX), 153:15–26.

White, F.M. 1984. *Heat Transfer*, Addison-Wesley, New York.

Wirtz, R.A. and McAuliffe, W. 1989. Experimental modeling of convective downstream from an electronic package row, *ASME J. of Electron. Packag.*. 11, 207–212.

Wirtz, R.A. and Dykshoorn, P. 1984. Heat transfer from arrays of flat packs in a channel flow, in *Proc. Inst. of Electron. Packag. Symp.*, 318–326.

INDEX

T - #0128 - 101024 - C0 - 234/156/14 [16] - CB - 9780849394478 - Gloss Lamination